AQA Physics

A LEVEL YEAR 1 AND AS

Jim Breithaupt

OXFORD

OXFORD

UNIVERSITY PRESS

Great Clarendon Street, Oxford, OX2 6DP, United Kingdom

Oxford University Press is a department of the University of Oxford.
It furthers the University's objective of excellence in research,
scholarship, and education by publishing worldwide. Oxford is a
registered trade mark of Oxford University Press in the UK and in
certain other countries

British Library Cataloguing in Publication Data
Data available

978 0 19 835188 7

10 9 8 7 6 5 4 3 2 1

Paper used in the production of this book is a natural, recyclable
product made from wood grown in sustainable forests.
The manufacturing process conforms to the environmental regulations
of the country of origin.

Printed in Great Britain by Bell and Bain Ltd, Glasgow

Artwork by Q2A Media

AS/A Level course structure

This book has been written to support students studying for AQA AS Physics and for students in their first year of studying for AQA A Level Physics. It covers the AS sections from the specification, the content of which will also be examined at A Level. The sections covered are shown in the contents list, which also shows you the page numbers for the main topics within each section. If you are studying for AS Physics, you will only need to know the content in the shaded box.

AS exam

A level exam

Year 1 content

1 Particles and radiation
2 Waves
3 Mechanics and energy
4 Electricity
5 Skills in AS Physics

Year 2 content

6 Further mechanics and thermal physics
7 Fields
8 Nuclear physics

Plus one option from the following:
- Astrophysics
- Medical physics
- Engineering physics
- Turning points in physics
- Electronics

A Level exams will cover content from Year 1 and Year 2 and will be at a higher demand.

Contents

How to use this book

This book contains many different features. Each feature is designed to support and develop the skills you will need for your examinations, as well as foster and stimulate your interest in physics.

Worked example

Step-by-step worked solutions.

Common misconception

Common student misunderstandings clarified.

Maths skill

A focus on maths skills.

Summary questions

1 These are short questions at the end of each topic.

2 They test your understanding of the topic and allow you to apply the knowledge and skills you have acquired.

3 The questions are ramped in order of difficulty. Lower-demand questions have a paler background, with the higher-demand questions having a darker background. Try to attempt every question you can, to help you achieve your best in the exams.

Specification references

→ At the beginning of each topic, there are specification references to allow you to monitor your progress.

Key term

Pulls out key terms for quick reference.

Synoptic link

These highlight how the sections relate to each other. Linking different areas of physics together becomes increasingly important, as many exam questions (particularly at A Level) will require you to bring together your knowledge from different areas.

Revision tip

Study tips contain prompts to help you with your revision. They can also support the development of your practical skills 🧪 and your mathematical skills √x̄.

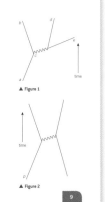

Practice questions at the end of each chapter, including questions that cover practical and maths skills.

1.1 Inside the atom

Specification reference: 3.2.1.1

The structure of an atom is shown in Figure 1.

- The **nucleus** contains most of the mass of the atom and its diameter is of the order of 10^{-5} times the diameter of a typical atom.
- The **atomic number Z** (or the proton number) of an atom is the number of protons in its nucleus.
- The **mass number A** (or the nucleon number) of an atom is the number of nucleons (i.e. protons and neutrons) in its nucleus.

Table 1 shows the charge and the mass of the proton, the neutron, and the electron relative to the charge and mass of the proton.

Isotopes

The atoms of an element each have the same number of protons but they can have different numbers of neutrons. Atoms of the same element with different numbers of neutrons are called **isotopes**. Figure 2 shows how we label them.

Specific charge

The **specific charge** of a charged particle is defined as its charge divided by its mass. The unit of specific charge is the coulomb per kilogram ($C\,kg^{-1}$).

> **Worked example**
> Calculate the specific charge of a nucleus of the carbon isotope $^{14}_{6}C$.
>
> *Solution*
>
> The charge of the nucleus $= 6e = 6 \times 1.60 \times 10^{-19}\,C = 9.60 \times 10^{-19}\,C$
>
> The mass of the nucleus $= 14 \times 1.67 \times 10^{-27} = 2.34 \times 10^{-26}\,kg$
>
> Specific charge $= \dfrac{9.60 \times 10^{-19}\,C}{2.34 \times 10^{-26}\,kg} = 4.10 \times 10^{7}\,C\,kg^{-1}$

We use the word **nucleon** for a proton or a neutron in the nucleus.

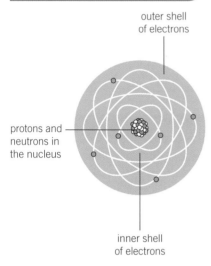

▲ **Figure 1** *Inside the atom*

▼ **Table 1** *Inside the atom*

	Charge relative to proton	Mass relative to proton
Proton	1	1
Neutron	0	1
Electron	−1	0.0005

Revision tip

charge of the proton $= e = 1.60 \times 10^{-19}\,C$

mass of the proton $= 1.67 \times 10^{-27}\,kg$

Key term

Isotopes are atoms of an element with different numbers of neutrons and the same number of protons.

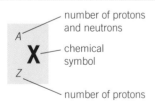

▲ **Figure 2** *Isotope notation*

Revision tip

Remember the charge of a nucleus is due to its proton number whereas the charge of an ion is due to how many electrons it has compared with an uncharged atom.

Summary questions

1   The table below gives some data about 4 different nuclei, A, B, C, and D.

	A	B	C	D
Atomic number	12	14	16	20
Mass number	24	30	32	39

    a How many neutrons and how many protons are there in D? *(1 mark)*

    b Which nucleus contains more neutrons than protons? *(1 mark)*

    c C is an isotope of sulfur (S). Another isotope E of sulfur has two fewer neutrons than C. Write down the isotope notation for E. *(1 mark)*

2   **a** State the number of protons and the number of neutrons in a nucleus of:

       i $^{17}_{8}O$ **ii** $^{9}_{4}Be$ *(2 marks)*

    b State and explain which of the two nuclei in **a** has the larger specific charge. *(2 marks)*

3   Calculate the specific charge of:

    a a uranium $^{238}_{92}U$ nucleus

    b a beryllium $^{9}_{4}Be$ ion with 2 electrons only. *(6 marks)*

1.2 Stable and unstable nuclei
Specification reference: 3.2.1.2

The strong nuclear force

The **strong nuclear force** keeps the nucleus together because it overcomes the electrostatic force of repulsion between the protons in the nucleus, and keeps the protons and neutrons together. Its key characteristics are:

- The specific size of its range, which is no more than about 3–4 fm.
- It is an attractive force from 3–4 fm to about 0.5 fm. At separations < 0.5 fm, it is a repulsive force that acts to prevent neutrons and protons being pushed into each other.
- It has the same effect between two protons as it does between two neutrons or a proton and a neutron.

Revision tip
Remember that 1 femtometre (fm) = 10^{-15} m

Radioactive decay

Naturally occurring radioactive isotopes release three types of radiation.

1 **Alpha radiation** consists of alpha particles which each comprise two protons and two neutrons. The symbol for an alpha particle is $_2^4\alpha$ because its proton number is 2 and its mass number is 4. When an unstable nucleus of an element X emits an alpha particle, its nucleon number A decreases by 4 and its atomic number Z decreases by 2. We can represent this change by means of the equation below.

$$_Z^A X \rightarrow\ _{Z-2}^{A-4}Y +\ _2^4\alpha$$

2 **Beta radiation** from naturally occurring isotopes consists of fast-moving electrons. The symbol for an electron as a beta particle is $_{-1}^{0}\beta$ (or β^-).

An unstable nucleus of an element X emits a β^- particle when a neutron in the nucleus changes into a proton. This type of change happens to nuclei that have too many neutrons. The beta particle is created when the change happens and is emitted instantly. In addition, an antiparticle with no charge, called an antineutrino (symbol $\bar{v}$), is emitted. We can represent this change by means of the equation below:

$$_Z^A X \rightarrow\ _{Z+1}^{A}Y +\ _{-1}^{0}\beta + \bar{v}$$

3 **Gamma radiation** (symbol γ) is electromagnetic radiation emitted by an unstable nucleus. It can pass through thick metal plates. It has no mass and no charge. It is emitted by a nucleus with too much energy, following an alpha or beta emission.

Revision tip
The neutrino and the antineutrino were 'hypothesised' to explain why beta particles from an isotope have a range of kinetic energies up to a maximum unlike alpha particles, which are emitted with a fixed proportion of the energy released. The existence of neutrinos and antineutrinos was confirmed about twenty years after the neutrino hypothesis was put forward.

Revision tip
There are two types of beta radiation. Nuclei that have too many protons emit beta radiation consisting of positrons. See Topic 1.4, Particles and antiparticles.

Summary questions

1 State two differences between the strong nuclear force and the electrostatic force. *(2 marks)*

2 An unstable nucleus $_Z^A X$ emits an α particle then a β^- particle to form a nucleus Y.
State:
  a the mass number of Y b the proton number of Y. *(2 marks)*

3 For the following radioactive decay equations, give the correct values for a, b, and c for each equation.
  a $_{29}^{64}Cu \rightarrow\ _b^a Zn +\ _{-1}^{c}\beta + \bar{v}$ *(2 marks)*
  b $_{90}^{228}Th \rightarrow\ _b^a Ra +\ _c^4\alpha$ *(2 marks)*

1.3 Photons

Electromagnetic waves

Light is just a small part of the spectrum of **electromagnetic waves**. In a vacuum, all *electromagnetic* waves travel at the speed of light, c, which is $3.00 \times 10^8 \, \text{m s}^{-1}$. The wavelength λ of **electromagnetic radiation** of frequency f in a vacuum is given by the equation

$$\lambda = \frac{c}{f}$$

▼ **Table 1** *The main parts of the electromagnetic spectrum*

Type	Radio	Microwave	Infrared	Visible	Ultraviolet	X-rays	Gamma rays
Wavelength range	>0.1 m	0.1 m–1 mm	1 mm–700 nm	700 nm–400 nm	400 nm–10 nm	10 nm–0.001 nm	<0.1 nm

Revision tip

The equation $E = hf$ gives the energy of a photon in joules. To convert joules into electron volts, remember

$$1 \, \text{eV} = 1.60 \times 10^{-19} \, \text{J}$$

Maths skill

Given the wavelength of a photon, you can calculate its energy in one step using the equation

$E = \dfrac{hc}{\lambda}$ (obtained by *substituting*

$f = \dfrac{c}{\lambda}$ into $E = hf$)

Synoptic link

You will revise photon theory in more detail in Topic 3.1, The photoelectric effect.

Photons

Electromagnetic waves are emitted as short bursts of waves, each burst leaving the source in a different direction. Each burst is a **photon**. The energy E of a photon depends on its frequency f in accordance with the equation

photon energy $E = hf$

where h is a constant referred to as the Planck constant. The value of h is $6.63 \times 10^{-34} \, \text{J s}$.

For a beam of photons of frequency f,

the power of the beam $= nhf$

where n is the number of photons in the beam passing a fixed point each second.

Worked example

Calculate the energy of a photon of wavelength 380 nm.

$h = 6.63 \times 10^{-34} \, \text{J s}, c = 3.00 \times 10^8 \, \text{m s}^{-1}$

Solution

To calculate the energy of a photon of this wavelength, we can use $E = \dfrac{hc}{\lambda}$

$$E = \frac{hc}{\lambda} = \frac{6.63 \times 10^{-34} \times 3.00 \times 10^8}{380 \times 10^{-9}} = 5.23 \times 10^{-19} \, \text{J}$$

Summary questions

$c = 3.00 \times 10^8 \, \text{m s}^{-1}, h = 6.63 \times 10^{-34} \, \text{J s}, e = 1.60 \times 10^{-19} \, \text{C}$

1. **a** State one similar property and one different property of a radio wave photon and a light photon. *(2 marks)*
   b Calculate the wavelength of an X-ray photon of energy 50 keV. *(2 marks)*

2. **a** Calculate the energy in J of a photon of wavelength 30 mm and state the part of the electromagnetic spectrum the photon is in. *(1 mark)*
   b Repeat part **a** for a photon of wavelength 3.0×10^{-10} m. *(2 marks)*

3. A 5.0 mW light-emitting diode emits light of wavelength approximately 600 nm. Calculate the maximum number of photons emitted per second by this LED. *(3 marks)*

1.4 Particles and antiparticles

Specification reference: 3.2.1.3

Rest energy

The equation $E = mc^2$ relates the total energy of a particle or antiparticle to its mass. When a particle is stationary, its rest mass (m_0) corresponds to **rest energy** m_0c^2 locked up as mass. Rest energy must be included in the conservation of energy.

The energy of a particle or antiparticle is often expressed in millions of electron volts (MeV), where **1 MeV = 1.60×10^{-13} J**.

Antimatter

For every type of particle there is a corresponding antiparticle that:

- annihilates the particle and itself if they meet, converting their total mass into photons

- has exactly the same rest mass as the particle

- has exactly opposite charge to the particle if the particle has a charge

- can be produced together with its corresponding particle in a process known as **pair production**. This can happen when a *single* photon with sufficient energy passing near a nucleus changes into a particle–antiparticle pair. The nucleus is necessary for conservation of momentum as well as conservation of energy.

Annihilation

- A particle and a corresponding antiparticle meet and their mass, including their rest mass, is all converted into radiation energy so that energy is conserved.

- Two photons of equal energy are produced in this process (as a single photon cannot ensure a total momentum of zero after the collision).

- If a particle and corresponding antiparticle each have rest energy E_0

    Minimum energy of each photon produced, $hf_{min} = E_0$

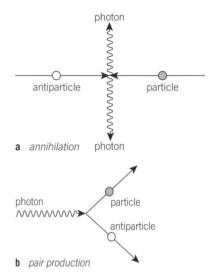

a *annihilation*

b *pair production*

▲ **Figure 1** *Particles and antiparticles*

Pair production

For a photon to produce a particle and antiparticle, each of rest energy E_0, the minimum energy of the photon must be equal to $2E_0$.

   Minimum energy of photon needed = $hf_{min} = 2E_0$

Worked example

The electron has a rest energy of 0.511 MeV. Calculate the minimum energy in J of a photon for pair production of an electron and a positron.

Solution

Minimum energy of the photon = $2E_0$ = 2 × 0.511 MeV = 1.022 MeV = 1.022 × 1.60×10^{-13} J = 1.64×10^{-13} J

Positron emission

The positron (symbol $_{+1}^{0}\beta$ or β^+) is the antiparticle of the electron, so it carries a positive charge. Positron emission takes place when a proton changes into a neutron in an unstable nucleus with too many protons. In addition, a neutrino (symbol $v_{(e)}$ for an electron neutrino), which is uncharged, is emitted.

$$_{Z}^{A}\text{X} \rightarrow {}_{Z-1}^{A}\text{Y} + {}_{+1}^{0}\beta + v$$

Revision tip

Notice in the positron emission equation that the numbers on each side

along the top add up to the same total (i.e., $A = A + 0$)

along the bottom add up to the same total (i.e., $Z = (Z - 1) + 1$)

Summary questions

1 MeV = 1.60×10^{-13} J

1 The rest energy of a proton is 938 MeV. Calculate the minimum energy of a photon for it to create a proton–antiproton pair. *(1 mark)*

2 The rest energy of an electron is 0.511 MeV. A positron created in an experiment has 0.250 MeV of kinetic energy. It collides with an electron at rest, creating two photons of equal energies as a result of annihilation.
   a Calculate the total energy of the positron and the electron. *(2 marks)*
   b Calculate the energy of each photon. *(1 mark)*

3 a Explain what is meant by pair production. *(1 mark)*
   b State two conditions necessary for a pair production event. *(1 mark)*

1.5 Particle interactions
Specification reference: 3.2.1.4

The electromagnetic force

This force acts between two charged objects due to the exchange of **virtual photons**. The photons are described as *virtual* because we can't detect them directly. The interaction is represented by the diagram shown in Figure 1. This is a simplified version of what is known as a **Feynman diagram**.

The weak nuclear force

The weak nuclear force is due to the exchange of particles referred to as **W bosons**. Unlike photons, these exchange particles:

- have a non-zero rest mass
- have a very short range of no more than about 0.001fm
- are positively charged (the W^+ boson) or negatively charged (the W^- boson).

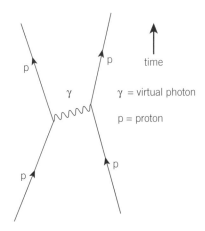

▲ **Figure 1** *Diagram for the electromagnetic force between two protons*

β decay

Figure 2 shows the diagram for each β decay process. Notice that:

- charge is conserved in both processes
- in β^- decay, a neutron changes into a proton and emits a W^- boson that decays into a β^- particle and an antineutrino
- in β^+ decay, a proton changes into a neutron and emits a W^+ boson that decays into a β^+ particle and a neutrino.

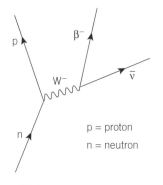

a β^- *decay*

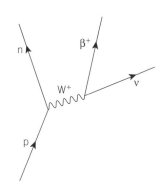

b β^+ *decay*

▲ **Figure 2** *W bosons in beta decay*

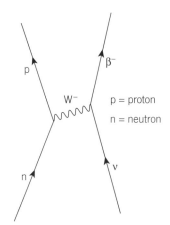

a *A neutron–neutrino interaction*

Neutrino interactions

Neutrinos and antineutrinos can interact with other particles. These rare interactions are due to W boson exchange. See Figure 3. Notice that:

- charge is conserved in both examples
- in **a**, a W^- boson is exchanged from a neutron to a neutrino so the neutron changes into a proton and the neutrino changes into a β^- particle (i.e., an electron)
- in **b**, a W^+ boson is exchanged from the proton to the antineutrino.so the proton changes into a neutron and the antineutrino changes into a β^+ particle (i.e., a positron).

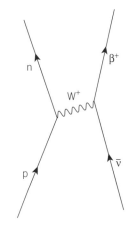

b *A proton–antineutrino interaction*

▲ **Figure 3** *The weak interaction*

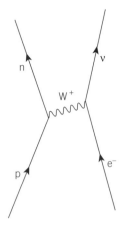

▲ **Figure 4** *Electron capture*

Revision tip

There are four fundamental forces in nature – the force of gravity, the electromagnetic force, the strong nuclear force, and the weak nuclear force. Remember that the photon and the W boson are the exchange particles of the electromagnetic force and the weak nuclear force respectively. The exchange particle of the strong nuclear force is the π meson. Scientists think the carrier of the force of gravity is the graviton – but it has yet to be observed!

Electron capture

Sometimes a proton in a proton-rich nucleus turns into a neutron as a result of interacting through the weak interaction with an inner-shell electron from outside the nucleus (**electron capture**). The W^+ boson emitted by the proton changes the electron into a neutrino.

The same overall change can happen when a proton and an electron collide at very high speed. In addition, for an electron with sufficient energy, the overall change could also occur as a W^- exchange from the electron to the proton.

Summary questions

$c = 3.00 \times 10^8 \, \mathrm{m \, s^{-1}}$

1 State the exchange particle involved when:
   a two electrons interact   (*1 mark*)
   b a neutron and a neutrino interact   (*1 mark*)
   c β^+ decay occurs.   (*1 mark*)

2 Sketch a diagram to represent:
   a β^- decay   (*2 marks*)
   b an interaction between a proton and an antineutrino in
     which the proton changes to a neutron and emits a W^+
     boson which then interacts with the antineutrino.   (*1 mark*)

3 a Sketch a diagram to represent electron capture.   (*2 marks*)
   b Describe the process of electron capture.   (*3 marks*)

Chapter 1 Practice questions

1 $_{11}^{23}$Na is a neutral atom of sodium.

  a How many of the following does this atom contain?

    i protons

    ii neutrons

    iii electrons. *(1 mark)*

  b Calculate the specific charge of an $_{11}^{23}$Na$^+$ ion. *(3 marks)*

2 A neutral chlorine (Cl) atom contains 38 nucleons.

  The specific charge of its nucleus is $4.30 \times 10^7 \, \text{C kg}^{-1}$.

  a Determine the number of protons in its nucleus. *(3 marks)*

  b The nucleus is unstable and decays by emitting a β$^-$ particle to become an argon (Ar) nucleus. Write down the equation for this decay. *(3 marks)*

  c Calculate the specific charge of the nucleus after it emits the β$^-$ particle. *(2 marks)*

3 The equation below represents the decay of a bismuth (Bi) isotope by the emission of an α particle to form an isotope of thallium (Th).

$$_{83}^{210}\text{Bi} \rightarrow {_b^a}\text{Th} + {_d^c}\alpha$$

  a Determine the values of *a, b, c,* and *d* shown in the equation. *(2 marks)*

  b State which type of force X is responsible for holding a stable nucleus together and what particles the force acts between.

  c State which type of force Y has to be overcome by X to hold the nucleus together and what particles Y acts between. *(2 marks)*

4 **a** A laser emits a narrow beam of light of wavelength 630 nm. Calculate the energy of a photon of wavelength 630 nm in electron volts. *(2 marks)*

  b The power of the beam of light is 4.0 mW. Calculate the number of photons per second emitted by the laser source. *(2 marks)*

5 **a** State two characteristics of the electromagnetic force. *(2 marks)*

  b The electromagnetic force acts between two objects due to the exchange of virtual photons. With the aid of a diagram, explain what is meant by this statement. *(2 marks)*

6 In a radioactive decay of a certain nucleus, a positron is emitted.

  a Explain what a positron is. *(1 mark)*

  b **i** Describe how the number of protons and the number of neutrons in a nucleus change when a positron is emitted. *(1 mark)*

    ii Name the fundamental force responsible for positron emission and name the exchange particle of this force. *(1 mark)*

    iii Figure 1 represents positron emission. Identify the particles or antiparticles represented by the letters *a, b, c, d,* and *e* on the diagram. *(3 marks)*

7 The equation below represents electron capture.

$$\text{p} + \text{e}^- \rightarrow \text{n} + \nu_e$$

  a Describe the process of electron capture. *(2 marks)*

  b Copy and complete Figure 2 to represent this process. *(2 marks)*

8 **a** Describe what happens in pair production and give *one* example of this process. *(3 marks)*

  b A particle of rest energy 0.51 MeV meets its corresponding antiparticle and they annihilate each other producing two photons as a result. Calculate the least possible energy of each photon produced in this process. *(2 marks)*

  c Discuss whether or not annihilation may be considered as pair production in reverse. *(3 marks)*

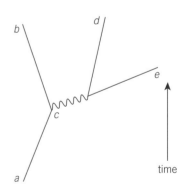

▲ Figure 1

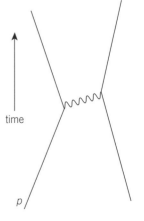

▲ Figure 2

2.1 The particle zoo

Specification reference: 3.2.1.5

▲ **Figure 1** *Creation and decay of a pion. The short spiral track is a π⁺ meson created when an antiproton from the bottom edge annihilates a proton. The π⁺ meson decays into an antimuon that spirals and decays into a positron.*

Synoptic link

For more about electron and muon neutrinos and strange particles, see Topics 2.3, Leptons at work and 2.4, Quarks and antiquarks.

Revision tip

Particles such as K mesons that are produced through the strong interaction and decay through the weak interaction producing leptons are called **strange particles**.

Synoptic link

Decays always obey the conservation rules for energy, momentum, and charge. See Topic 2.5, Conservation rules.

Muons and mesons

Cosmic rays are high-energy particles from stars including the Sun. When cosmic rays enter the Earth's atmosphere, they collide with atoms creating photons and new short-lived particles and antiparticles, including:

- the **muon** or heavy electron (symbol μ), a negatively charged particle with a rest mass over 200 times the rest mass of the electron
- the **pion** or π **meson**, a particle which can be positively charged (π⁺), negatively charged (π⁻), or neutral (π⁰), and has a rest mass greater than a muon but less than a proton
- the **kaon** or **K meson**, which also can be positively charged (K⁺), negatively charged (K⁻), or neutral (K⁰), and has a rest mass greater than a pion but still less than a proton.

Decay modes

The particles listed above can also be created using accelerators in which protons collide head-on with other protons at high speed. The kinetic energy of the protons is converted into mass in the creation of these new particles.

K mesons and π mesons are produced through the strong interaction. However, the decay of K mesons takes longer than expected and produces π mesons which means they must decay via the weak interaction.

The rest masses, charge (if they were charged), and lifetimes of the new particles have been measured. Their antiparticles, including the **antimuon**, have been detected. Their decay modes are:

- A K meson can decay into π mesons, or a muon and an antineutrino, or an antimuon and a neutrino.
- A charged π meson can decay into a muon and an antineutrino, or an antimuon and a neutrino. A π⁰ meson decays into high-energy photons.
- A muon decays into a muon neutrino, an electron, and an electron antineutrino. An antimuon decays into a muon antineutrino, a positron, and an electron neutrino.

Summary questions

1   a   Which of the particles below are negatively charged?   *(1 mark)*
    b   Which one of the particles below is uncharged and unstable?
        electron   K⁰ meson   antimuon   neutron
        π⁺ meson   antiproton   *(1 mark)*

2   State whether the strong or the weak interaction acts when:
    a   a muon decays   *(1 mark)*
    b   a K⁰ meson is produced   *(1 mark)*
    c   a π⁺ meson decays.   *(1 mark)*

3   a   A muon can travel much further than a π meson before they decay. What does this tell you about a muon compared with a π meson?   *(1 mark)*
    b   In terms of their properties, state one similarity and one difference between a muon and a π⁻ meson.   *(2 marks)*

2.2 Particle sorting
Specification reference: 3.2.1.5

Classifying particles and antiparticles

All the particles we have discussed so far except exchange particles are classified as **hadrons** and **leptons**, according to whether or not they interact through the strong interaction.

Hadrons are further classified into two groups:

1 **Baryons** are protons and all other hadrons (including neutrons) that decay into protons, either directly or indirectly.

2 **Mesons** are hadrons that do *not* include protons in their decay products. In other words, kaons and pions are not baryons.

Energy matters

When a particle or antiparticle collides with another particle or antiparticle:

$$\begin{matrix} \text{the rest energy of} \\ \text{the products} \end{matrix} = \begin{matrix} \text{total energy} \\ \text{before} \end{matrix} - \begin{matrix} \text{the kinetic energy} \\ \text{of the products} \end{matrix}$$

> ### Synoptic link
>
> **Conservation of momentum**
>
> Momentum is always conserved in a collision. *If the particles collide head-on with equal and opposite momentum*, their total momentum is zero. So all the total energy is available to create new particles and antiparticles. See Topic 9.3, Conservation of momentum, for more information.

> ### Key term
>
> **Leptons** are particles and antiparticles that do not interact through the strong interaction (e.g., electrons, muons, neutrinos, and their antiparticles). They can interact through the weak interaction, the gravitational interaction, and through the electromagnetic interaction (if charged).

> ### Key term
>
> **Hadrons** are particles and antiparticles that can interact through the strong interaction (e.g., protons, neutrons, π mesons, K mesons, and their antiparticles). They can also interact through the force of gravity, the weak interaction, and the electromagnetic interaction (if charged). Apart from the proton, which is stable, hadrons tend to decay through the weak interaction.

> ### Synoptic link
>
> Baryons and mesons are composed of smaller particles called **quarks** and **antiquarks**. See Topic 2.4, Quarks and antiquarks.

> ### Revision tip
>
> Rest energies are usually expressed in MeV ($= 10^6$ eV) or GeV ($= 1000$ MeV). For example, the rest energy of the proton is 938 MeV or 0.938 GeV.

> ### Summary questions
>
> 1 a i What property distinguishes a hadron from a lepton? *(1 mark)*
>   ii What property distinguishes a baryon from a meson? *(1 mark)*
>   b State whether each of the following particles or antiparticles is a baryon, a meson, or a lepton:
>   i a π meson   ii an antiproton   iii a neutrino. *(2 marks)*
>
> 2 A K^+ meson can decay into two π mesons.
>   a Complete the following equation for this decay: $K^+ \longrightarrow$ *(1 mark)*
>   b Use the rest energy values below to calculate the maximum kinetic energy of the π mesons, assuming the K^+ meson is at rest before it decays. *(2 marks)*
>   Rest energies : K^+ = 494 MeV, π^+ and π^- = 140 MeV, π^0 = 135 MeV
>
> 3 Two protons, X and Y, moving in opposite directions each with 1.00 GeV of kinetic energy collide and produce a further proton and an antiproton.
>   a Calculate the kinetic energy of the 3 protons and the antiproton after the collision. *(3 marks)*
>   b Explain why a further proton and antiproton could not be produced if X or Y was at rest and the total initial kinetic energy was still 2.00 GeV. *(2 marks)*
>   Rest energies: proton and antiproton = 0.94 GeV

2.3 Leptons at work

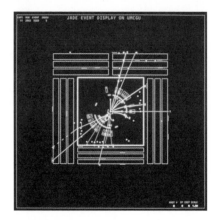

▲ **Figure 1** *A two-jet event after an electron–positron collision*

All the experiments on leptons indicate they don't break down into non-leptons. A lepton can change into other leptons by emitting or absorbing a W boson. They can also interact in high-energy collisions to produce hadrons.

Neutrino types

Neutrinos and antineutrinos produced in beta decays are different from those produced by muon decays. We use the symbol v_μ for the **muon neutrino** and v_e for the **electron neutrino** (and similarly for the two types of antineutrinos).

Lepton rules

Some lepton changes are possible and some changes are never observed even though charge is conserved. To explain this, we need to assign +1 to a lepton, −1 to an antilepton, and 0 for any non-lepton and require lepton numbers to be conserved separately for muon lepton (and for electron leptons).

1   When an electron neutrino interacts with a neutron, it can produce a proton and an electron: $v_e + n \rightarrow p + e^-$

Electron lepton numbers are conserved: $1 + 0 = 0 + 1$ ✓
BUT the following change is not possible even though charge is conserved.
$$v_e + n \nrightarrow \bar{p} + e^+$$
Electron lepton numbers are not conserved: $1 + 0 \neq 0 + -1$ ✗

2   In muon decay, the muon changes into a muon neutrino. In addition, an electron is created to conserve charge and a corresponding antineutrino is created to conserve lepton number. For example: $\mu^- \rightarrow e^- + \bar{v}_e + v_\mu$

Electron lepton numbers are conserved: $0 = +1 - 1 + 0$ ✓
Muon lepton numbers are conserved: $+1 = 0 + 0 + 1$ ✓

Note: a muon *cannot* decay into a muon antineutrino, an electron, and an electron antineutrino even though charge is conserved. This is because the muon lepton number is not conserved. $\mu^- \rightarrow e^- + \bar{v}_e + v_\mu$

Electron lepton numbers are conserved: $0 = 1 - 1 + 0$ ✓
Muon lepton numbers are *not* conserved: $+1 \neq 0 + 0 - 1$ ✗

Common misconception

Always consider charge conservation first because the individual charges are easy to remember. Also, when considering muons and antimuons, remember the muon is a particle not an antiparticle.

▼ **Table 1**

	Charge Q	Lepton number L
Electron	−1	+1
Muon		
Positron		
Muon antineutrino		
Antiproton		

Summary questions

1   Complete Table 1 to show the relative charge and lepton number of each particle or antiparticle in the table. *(3 marks)*

2   A muon decays into an electron, a neutrino, and an antineutrino.
   a   Complete the equation below representing this decay: *(2 marks)*
   $$\mu^- \rightarrow \quad + \quad v_\mu +$$
   b   Show that electron lepton numbers and muon lepton numbers are conserved in the above equation. *(2 marks)*

3   State whether or not each of the following reactions is permitted, giving a reason if it is not permitted:
   a $\bar{v}_e + p \rightarrow n + e^-$ *(2 marks)*
   b $v_e + p \rightarrow n + e^+$ *(2 marks)*
   c $e + p \rightarrow n + \bar{v}_e$ *(2 marks)*
   d $\mu^+ \rightarrow \bar{v}_e + e^+ + \bar{v}_\mu$ *(2 marks)*

2.4 Quarks and antiquarks

Specification reference: 3.2.1.6; 3.2.1.7

Strangeness

Strange particles such as K mesons are produced in pairs by making π mesons collide with protons and neutrons. One of the pair (the K meson) always decays into π mesons only whereas the other one always decays into π mesons and protons.

To explain the interactions and decays of strange particles,

- a **strangeness number** S was introduced for each particle and antiparticle (starting with +1 for the K$^+$ meson) so that strangeness is always conserved in strong interactions.
- non-strange particles (i.e., the proton, the neutron, pions, leptons) were assigned zero strangeness.

Strangeness is always conserved in a strong interaction, whereas strangeness can change by 0, +1, or −1 in weak interactions.

Look at the reactions in Table 1.

Reaction 1 is observed: the initial strangeness is zero so the Σ^0 strangeness must be −1 as the K$^+$ strangeness by definition is +1.

Reaction 2 is not observed: the initial strangeness is zero and the Σ^0 strangeness is −1. So the K$^-$ strangeness cannot be +1.

▼ **Table 1** *Some reactions that were predicted*

1	observed	$\pi^+ + N \longrightarrow K^+ + \Sigma^0$
2	*not* observed	$\pi^- + N \longrightarrow K^- + \Sigma^0$

The quark model

The properties of the hadrons, such as charge, strangeness, and rest mass can be explained by assuming they are composed of smaller particles known as quarks and antiquarks. The three types of quarks and antiquarks you need to study in this course are: the up, down, and strange quarks (u, d, s) and their antiquarks ($\bar{u}, \bar{d}, \bar{s}$). The charge Q, strangeness S, and baryon number B of the u, s, and d quarks are shown in Table 2. Their corresponding values for their antiquarks have the opposite sign (Q for $\bar{u} = -\frac{2}{3}$).

> **Revision tip**
> You may be asked to work out the charge or strangeness of any baryon or antibaryon from its quark composition.

▼ **Table 2** *Quark properties*

	u	d	s
Q	$+\frac{2}{3}$	$-\frac{1}{3}$	$-\frac{1}{3}$
S	0	0	−1
B	$+\frac{1}{3}$	$+\frac{1}{3}$	$+\frac{1}{3}$

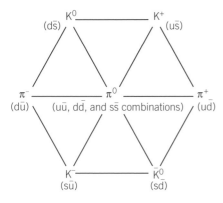

▲ **Figure 1** *Quark combinations for the mesons*

Quark combinations

The rules for combining quarks to form baryons and mesons are astonishingly simple.

A **meson** consists of a quark and an antiquark. Figure 1 shows all nine different quark–antiquark combinations and the meson in each case. Notice that:

- each pair of charged mesons (e.g., π^+ and π^- or K^+ and K^-) is a particle–antiparticle pair
- each type of K meson contains one strange quark or antiquark.

Baryons and **antibaryons** are hadrons that consist of three quarks for a baryon or three antiquarks for an antibaryon.

- A proton is the uud combination, an antiproton is the $\overline{u}\,\overline{u}\,\overline{d}$ combination.
- A neutron is the udd combination.
- The Σ particle is a baryon containing a strange quark.

The proton is the only stable baryon. A free neutron decays into a proton, releasing an electron and an electron antineutrino, as in β^- decay.

Quarks and beta decay

In β^- decay, a down quark in a neutron changes to an up quark. See Figure 2a.

In β^+ decay, an up quark in a proton changes to a down quark. See Figure 2b.

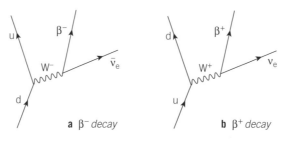

▲ **Figure 2** *Quark changes in beta decay*

Summary questions

1 Determine the quark composition and strangeness of each of these hadrons:
  a **i** a π^+ meson *(1 mark)*
    ii an antineutron *(1 mark)*
    iii a K^+ meson. *(1 mark)*
  b Show that the antiparticle of a K^- meson is a K^+ meson. *(2 marks)*

2 **a** In terms of quarks, draw a diagram to represent β^- decay. *(2 marks)*
  b Describe the changes that are represented in your diagram. *(3 marks)*

3 **a** A Σ^- particle is a baryon which has a strangeness of -1. Determine its quark composition. *(3 marks)*
  b Show that strangeness is conserved in the reaction below. *(3 marks)*
$$\pi^- + p \rightarrow K^+ + \Sigma^-$$

2.5 Conservation rules
Specification reference: 3.2.1.6; 3.2.1.7

Conservation of energy, conservation of momentum, and conservation of charge apply to all changes in science. Particles and antiparticles obey further conservation rules when they interact. These further rules are based on what reactions are observed and what reactions are not observed.

Conservation of leptons numbers: The total lepton number for each lepton branch before the change is equal to the total lepton number for that branch after the change.

Conservation of strangeness: In any strong interaction, strangeness is always conserved.

In addition, we know from observation that meson numbers are not conserved and baryon numbers are conserved if we assign a baryon number of:

- +1 to any baryon and −1 to any antibaryon
- 0 to any meson or lepton.

Quarks, antiquarks, and their baryon numbers

We can apply the baryon conservation rule to quarks, antiquarks, and leptons by assigning $+\frac{1}{3}$ to any quark, $-\frac{1}{3}$ to any antiquark, and 0 to any lepton.

The first reaction in Table 1 is shown in Figure 1 in terms of quarks and by the equation below. The baryon numbers under the equation show that the total baryon number is the same on each side.

$$u\,u\,d + \overline{u}\,\overline{u}\,\overline{d} \rightarrow u\,\overline{d} + \overline{u}\,d$$
$$\left(+\frac{1}{3}+\frac{1}{3}+\frac{1}{3}\right) + \left(-\frac{1}{3}-\frac{1}{3}-\frac{1}{3}\right) = \left(+\frac{1}{3}-\frac{1}{3}\right) + \left(+\frac{1}{3}-\frac{1}{3}\right)$$

The s quark decays through the weak interaction by changing into a u quark and emitting a W^- boson which can then decay into an electron and an antineutrino.

▼ **Table 1** *Conservation of baryon numbers*

Example 1	observed
Reaction	$p + \overline{p} \rightarrow \pi^+ + \pi^-$
Baryon numbers	$1 - 1 = 0 + 0$

Example 2	not observed
Reaction	$p + \overline{p} \rightarrow p + \pi^-$
Baryon numbers	$1 - 1 \neq 1 + 0$

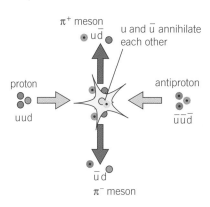

▲ **Figure 1** *Using the quark model*

Summary questions

1. The following reaction was observed when two protons collided head-on at the same speed:
$$p + p \rightarrow p + p + p + \overline{p}$$
   a. Show that charge and baryon number are conserved. *(2 marks)*
   b. The rest energy of the proton is 0.94 GeV. Calculate the minimum kinetic energy each of the initial protons must have to create a proton and an antiproton. *(2 marks)*

2. a. State the baryon number of:
      i. an up antiquark   ii. a positron   iii. a neutron. *(2 marks)*
   b. State the lepton number of a:
      i. muon   ii. muon antineutrino   iii. K^- meson. *(2 marks)*

3. The lambda particle $\wedge^0$ is a baryon which has a strangeness of −1.
   a. Determine its quark composition. Explain your method. *(3 marks)*
   b. A $\wedge^0$ particle can be produced together with another particle X in the following reaction. $\pi^- + p \rightarrow \wedge^0 + X$
   By considering the appropriate conservation rules, identify particle X. *(4 marks)*

Chapter 2 Practice questions

1 $\bar{p}$ e^+ K^- v_e μ^+

From the list of particles, identify the:

a hadrons

b leptons

c antibaryons

d mesons.   (*2 marks*)

2 **a** State the distinction between hadrons and leptons.   (*1 mark*)

b A positive muon may decay in the following way:

$$\mu^+ \rightarrow e^+ + v_e + \bar{v}_\mu$$

  i State the type of interaction that acts in the above decay process.

  ii Show that conservation of leptons applies to the above equation.

  iii Give *one* difference and *one* similarity between a negative muon and an electron.   (*3 marks*)

3 **a** The three-quark model originated from a theory that explained the properties of baryons and mesons. State the quark composition of a neutron and explain why the neutron is uncharged.   (*2 marks*)

b Show that in the three-quark model, the antiparticle of a non-strange charged meson is another non-strange charged meson.   (*2 marks*)

4 **a** A particle is made up of an up quark and a down antiquark.

  i Name the particle that has this type of structure.

  ii Explain its charge in terms of quarks.   (*2 marks*)

b A negatively charged baryon X has a charge of −1, a baryon number of 1, and a strangeness of −2.

  i Deduce its quark composition and explain your reasoning.

  ii Which baryon will this particle eventually decay into?   (*5 marks*)

5 **a** With the aid of a diagram, describe the changes that take place in terms of quarks when a β^+ particle is emitted from a nucleus.   (*3 marks*)

b A suggested decay for the positive muon (μ^+) is

$$\mu^+ \rightarrow e^+ + \bar{v}_e + v_\mu$$

Showing your reasoning clearly, deduce whether this decay satisfies the conservation rules that relate to baryon number, lepton number, and charge.   (*3 marks*)

6 The following equation represents the collision of a K^- meson with a proton, resulting in the production of a baryon X and a π^- meson:

$$K^- + p \rightarrow \pi^- + X$$

a The K^- has a strangeness of −1. Deduce the strangeness S, charge Q, and baryon number B of X.   (*3 marks*)

b Discuss whether or not X and a π meson could be produced in a collision between a π^- meson and a proton.   (*2 marks*)

7 State one similarity and one difference between:

a an antimuon and a proton

b a π^0 meson and a K^0 meson

c a π^+ meson and a neutron.   (*3 marks*)

8 Determine the charge Q and the strangeness S of:

a a uss baryon

b a uds baryon

c a $\overline{dds}$ baryon

d a $u\bar{s}$ meson.   (*4 marks*)

3.1 The photoelectric effect

Specification reference: 3.2.2.1

The **photoelectric effect** is the emission of electrons from a metal surface when electromagnetic radiation above a certain frequency is directed at the surface. The electrons emitted in this process are called **photoelectrons**.

The main observations about the photoelectric effect are listed below:

1 Photoelectric emission of electrons from a metal surface does not take place if the frequency of the incident electromagnetic radiation is less than a minimum value known as the **threshold frequency f_{min}**. This minimum frequency depends on the type of metal.

2 The number of electrons emitted per second is proportional to the intensity of the incident radiation, provided its frequency $f > fmin$.

3 Photoelectric emission occurs instantly when incident radiation is directed at the surface and regardless of intensity.

Einstein's explanation of the photoelectric effect

The wave theory of light cannot explain either the existence of a threshold frequency or why photoelectric emission occurs without delay. Einstein explained the photoelectric effect by assuming that:

- light is composed of wavepackets or **photons**, each of energy equal to hf, where f is the frequency of the light and h is the Planck constant

- an electron at or near the metal surface absorbs a *single* photon from the incident light and so gains energy equal to hf, (the photon's energy)

- an electron can leave the metal surface if the energy gained from a single photon is equal to, or greater than, the **work function ϕ** of the metal.

The *maximum* kinetic energy of an emitted electron is therefore $E_{Kmax} = hf - \phi$ so $E_{Kmax} > 0$.
Rearranging this equation gives $hf = E_{Kmax} + \phi$

Emission can take place from a metal surface provided $hf > \phi$. Therefore, threshold frequency $f_{min} = \dfrac{\phi}{h}$.

Stopping potential

The minimum potential needed to stop photoelectric emission from a metal is called the **stopping potential V_s**. At this potential, the maximum kinetic energy E_{Kmax} of an emitted electron is reduced by an amount equal to eV_s to zero. Hence $E_{Kmax} = eV_s$.

Summary questions

$h = 6.63 \times 10^{-34}$ J s, $c = 3.00 \times 10^8$ m s^{-1}, $e = 1.60 \times 10^{-19}$ C

1 What is meant by the work function of a metal? *(1 mark)*

2 Explain why photoelectric emission from a metal surface only takes place if the frequency of the incident radiation is greater than or equal to a certain value. *(3 marks)*

3 A metal surface at zero potential emits electrons from its surface if light of wavelength less than or equal to 390 nm is directed at it.
   a Determine the work function of the metal. *(2 marks)*
   b Explain why light of wavelength greater than 390 nm will not cause photoelectric emission from this metal. *(2 marks)*

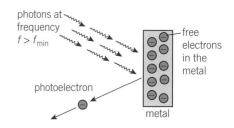

▲ **Figure 1** *Explaining the photoelectric effect – one electron absorbs one photon*

photons at frequency $f > f_{min}$

free electrons in the metal

photoelectron

metal

3.2 More about photoelectricity

Specification reference: 3.2.2.1

Synoptic link

If n electrons are emitted in time t, the charge flow Q $(= It) = ne$. Therefore, the number of electrons emitted per second, $\frac{n}{t} = \frac{I}{e}$.
See Topic 12.1, Current and charge, for $Q = It$.

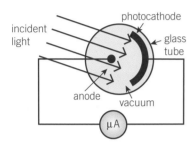

▲ **Figure 1** *Using a vacuum photocell*

Maths skill

A graph of E_{Kmax} against f is a straight line of the form $y = mx + c$. This is in accordance with the equation $E_{Kmax} = hf - \phi$ as $y = E_K$ and $x = f$.

Note that:

- the gradient of the line $m = h$ and the y-intercept $c = -\varphi$

- the x-intercept is equal to the threshold frequency.

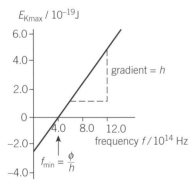

▲ **Figure 2** *A graph of E_{Kmax} against frequency*

$$\sqrt{v_x^2 + v_y^2} = \sqrt{16^2 + 30.4^2}$$

More about conduction electrons

The conduction electrons in a metal move about at random. The work function ϕ of a metal is much greater than the average kinetic energy of a conduction electron in a metal at 300 K.

- When a conduction electron absorbs a photon of energy hf, its kinetic energy increases by hf.

- If the energy of the photon $hf \geq \phi$, the conduction electron can leave the metal unless it collides with other electrons and positive ions and loses its extra kinetic energy.

The vacuum photocell

The photocell in Figure 1 contains a metal plate (the photocathode) and a smaller metal electrode, the anode. When photoelectric emission takes place, electrons are emitted from the cathode and are attracted to the anode. The microammeter in the circuit is used to measure the photoelectric current, which is proportional to the number of electrons per second that transfer from the cathode to the anode.

The photoelectric current is proportional to the intensity of the light incident on the cathode because:

- the light intensity is proportional to the number of photons per second incident on the cathode

- each photoelectron must have absorbed one photon to escape from the metal surface.

Summary questions

$h = 6.63 \times 10^{-34}$ J s, $c = 3.00 \times 10^8$ m s^{-1}, $e = 1.60 \times 10^{-19}$ C

1 Explain what is meant by the stopping potential of a metal. *(1 mark)*

2 A vacuum photocell is connected to a microammeter. When light is directed at the cathode of the photocell, the microammeter reads 0.85 μA.
   a i Explain why the microammeter gives a non-zero reading when light is directed at its cathode. *(2 marks)*
      ii Calculate the number of photoelectrons emitted per second by the photocathode of the photocell.
         Hint: Remember $Q = It$. *(2 marks)*
   b The intensity of the incident light is gradually reduced to zero. Describe and explain how the microammeter reading changes as a result. *(3 marks)*

3 The cathode of a photocell has a work function of 1.30 eV. A beam of light of wavelength 430 nm is directed at the photocathode of a vacuum photocell. Calculate:
   a the energy of a single light photon of this wavelength in eV *(1 mark)*
   b the maximum kinetic energy of a photoelectron in eV. *(2 marks)*

3.3 Collisions of electrons with atoms

Specification reference: 3.2.2.2

Ionisation

An **ion** is a charged atom. Adding electrons to an uncharged atom makes the atom into a negative ion. Removing electrons from an uncharged atom makes the atom into a positive ion (see Figure 1).

The electron volt and potential difference

The **electron volt** is a unit of energy which is defined as equal to the work done when an electron is moved through a potential difference (pd) of 1 V. For a charge q moved through a pd V, the work done $= qV$. Therefore, the work done when an electron moves through a potential difference V is equal to $e \times V$.

Excitation by collision

Gas atoms can absorb energy from colliding electrons without being ionised. This process, known as **excitation**, happens at certain values of energy, which are characteristic of the atoms of the gas. If a colliding electron:

- loses all its kinetic energy when it causes excitation, the current due to the flow of electrons through the gas is reduced

- does not have enough kinetic energy to cause excitation, it is deflected by the atom, with no overall loss of kinetic energy.

When excitation occurs, the colliding electron makes an electron inside the atom move from an inner shell to an outer shell (see Figure 2). Energy is needed for this process, because the atomic electron moves away from the nucleus of the atom. The excitation energy is always less than the ionisation energy of the atom, because the atomic electron is not removed completely from the atom when excitation occurs.

Key term

Any process of creating ions is called **ionisation**.

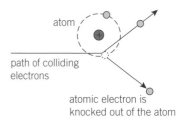

▲ **Figure 1** *Ionisation by collision*

Synoptic link

You met the electron volt briefly in Topic 2.1, The particle zoo.

Revision tip

Don't confuse atomic electrons with electrons that hit the atom.

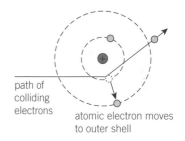

▲ **Figure 2** *A simple model of excitation by collision*

Summary questions

$e = 1.60 \times 10^{-19}$ C

1   Explain the difference between ionisation and excitation. *(2 marks)*

2   a   The hydrogen atom has an ionisation energy of 13.6 eV. Calculate this ionisation energy in joules. *(1 mark)*

    b   An electron with 16.2 eV of kinetic energy collides with a hydrogen atom and ionises it. Calculate the kinetic energy, in eV, of the electron after the collision if the collision causes an electron in the atom to be emitted from the atom with 1.0 eV of kinetic energy. *(2 marks)*

3   a   Describe what happens to a gas atom when an electron from outside the atom collides with it and causes it to absorb energy from the electron without being ionised. *(2 marks)*

    b   Explain why a gas atom cannot absorb energy from a slow-moving electron that collides with it. *(1 mark)*

3.4 Energy levels in atoms

Specification reference: 3.2.2.3

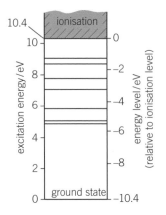

▲ **Figure 1** *The energy levels of the mercury atom*

Revision tip

The ionisation level may be considered as the zero reference level for energy, instead of the ground state level. The energy levels below the ionisation level would then need to be shown as negative values, as shown on the right-hand side of Figure 1.

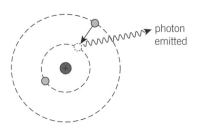

▲ **Figure 2** *De-excitation by photon emission*

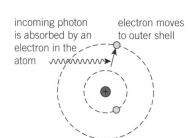

▲ **Figure 3** *Excitation by photon absorption*

Electrons in atoms

Figure 1 is an energy level diagram for an atom showing the allowed energy values of the atom. Each allowed energy corresponds to a certain electron configuration in the electron shells of the atom. The ground state of the atom is its lowest energy state. When an atom in the ground state absorbs energy, one of its electrons moves to a higher energy level so the atom becomes excited. The ionisation energy is the energy needed to ionise the atom from its ground state.

De-excitation

An excited atom is unstable because an electron that moves to an outer shell leaves a vacancy in the shell it moves from. The atom de-excites when the vacancy is filled by an electron from an outer shell. When this happens, the electron moves to a lower energy level and emits a photon (see Figure 2). When an electron moves from energy level E_1 to a lower energy level E_2, **the energy of the emitted photon, $hf = E_1 - E_2$**

Excitation using photons

An electron in an atom can absorb a photon and move to an outer shell where a vacancy exists – but only if the energy of the photon is *exactly* equal to the difference between the final and initial energy levels of the atom (see Figure 3).

Fluorescence

Certain substances **fluoresce** or glow with visible light when they absorb ultraviolet radiation. This is what happens in a **fluorescent tube**. The **tube** is a glass tube with a fluorescent coating on its inner surface. The tube contains mercury vapour at low pressure. When the tube is on, it emits visible light

- because the mercury atoms collide with each other and with electrons in the tube and become excited
- the mercury atoms emit ultraviolet photons (and photons with less energy) when they de-excite
- the ultraviolet photons are absorbed by the atoms of the fluorescent coating, causing excitation of the atoms
- the coating atoms de-excite in steps and emit visible photons.

Summary questions

$e = 1.60 \times 10^{-19}$ C

1 Figure 1 shows some of the energy levels of the mercury atom.
   a Estimate the energy needed to excite the atom from the ground state to the 4th excited level above the ground state. *(1 mark)*
   b Mercury atoms in this excited state can de-excite directly or indirectly to the ground state.
      i State and explain how many different photon energies are possible from this excited state. *(3 marks)*
      ii Which de-excitation from this state releases the highest photon energy? *(1 mark)*

3.5 Energy levels and spectra

Specification reference: 3.2.2.3

A continuous spectrum has a continuous range of wavelengths. For example, the spectrum of light from a filament lamp covers the visible spectrum from deep violet at less than 400 nm to deep red at about 650 nm.

A line emission spectrum has discrete lines of different wavelengths. The wavelengths of the lines of a line spectrum of an element are characteristic of the atoms of that element. This is because the energy levels of each type of atom are unique to that atom.

- Each line in a line spectrum is due to light of a certain colour and therefore a certain wavelength.

- The photons that produce a line all have the same energy, which is different from the energy of the photons that produce any other line.

- Each photon is emitted when an atom de-excites due to one of its electrons moving to an inner shell.

- If the electron moves from energy level E_1 to a lower energy level E_2.

the energy of the emitted photon $hf = E_1 - E_2$

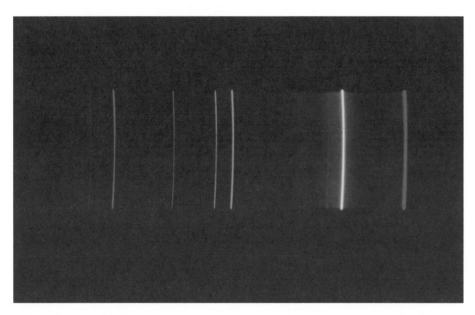

▲ **Figure 1** *A line spectrum*

Summary questions

$h = 6.63 \times 10^{-34}$ J s, $c = 3.00 \times 10^8$ m s^{-1}, $e = 1.60 \times 10^{-19}$ C

1 Describe what a line emission spectrum is and explain how it is produced. *(3 marks)*

2 The line spectrum of hydrogen includes two lines at 656 nm and 486 nm.
   a Calculate the energy of a photon of wavelength 656 nm. *(1 mark)*
   b The lines are due to electron transitions from two different energy levels X and Y to the same lower level. Level X is higher than level Y. Calculate the wavelength of a photon emitted by an electron that moves from X to Y.
   Hint: The first step is to calculate the energy of the 486 nm photon.
   (3 marks)

3.6 Wave–particle duality

Specification reference: 3.2.2.4

Synoptic link

See Topic 5.6, Diffraction, for more about diffraction and Topic 3.1, The photoelectric effect, for more about photoelectricity.

Key term

The **momentum** of a particle is defined as its mass multiplied by its velocity.

Common misconception

Don't mix up matter waves and electromagnetic waves and don't confuse their equations.

The dual nature of light

1 The *wave-like nature of light* is observed when **diffraction** of light takes place. This happens, for example, when light passes through a narrow slit.

2 The *particle-like nature of light* is observed in the photoelectric effect.

The dual nature of matter

1 The *particle-like nature of matter* is observed, for example, when electrons in a beam are deflected by a magnetic field.

2 The hypothesis that matter particles also have a *wave-like nature* was put forward by Louis de Broglie in 1923. His theory was that matter particles have a dual wave–particle nature and that they have a wavelength that depends on their momentum, p, according to the equation

$$\lambda = \frac{h}{p}$$

Evidence for de Broglie's hypothesis

A narrow beam of electrons can be diffracted by a metal foil as shown in Figure 1.

- The rows of atoms in each tiny crystal in the metal cause the electrons in the beam to be diffracted. The electrons are diffracted in certain directions only to form rings on a fluorescent screen at the end of the tube.

- Increasing the speed of these electrons makes their de Broglie wavelength smaller. So less diffraction occurs and the rings become smaller.

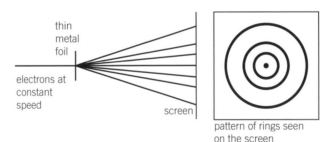

▲ **Figure 1** *Diffraction of electrons*

Summary questions

$h = 6.6 \times 10^{-34}$ J s, the rest mass of an electron = 9.11×10^{-31} kg, the rest mass of a proton = 1.67×10^{-27} kg

1 Explain what is meant by the dual wave–particle nature of matter particles. *(2 marks)*

2 **a** Calculate the de Broglie wavelength of an electron moving at a speed of 2.0×10^7 m s^{-1}. *(2 marks)*

   b Calculate the speed at which a proton would have the same de Broglie wavelength as the electron in **a**. *(2 marks)*

Chapter 3 Practice questions

$h = 6.63 \times 10^{-34} \,\text{J s}$, $e = 1.60 \times 10^{-19} \,\text{C}$, $c = 3.00 \times 10^{8} \,\text{m s}^{-1}$

rest mass of an electron = $9.11 \times 10^{-31} \,\text{kg}$

rest mass of a proton = $1.67 \times 10^{-27} \,\text{kg}$

1 a Describe the photoelectric effect. *(1 mark)*

   b State and explain why the photoelectric effect cannot be observed with a particular metal if the frequency of the incident radiation is too small. *(4 marks)*

2 The threshold frequency of light incident on a certain metal surface is $2.9 \times 10^{14} \,\text{Hz}$.

   a Explain what is meant by threshold frequency. *(1 mark)*

   b Calculate the maximum kinetic energy of the photoelectrons emitted from this metal surface when light of wavelength 560 nm is directed at the metal surface. Give your answer to an appropriate number of significant figures. *(3 marks)*

   c Sketch a graph on the axes shown in Figure 1 to show how the maximum kinetic energy $E_{K\text{max}}$ of the photoelectrons emitted from a metal surface varies with the frequency f of the incident light. Indicate the threshold frequency f_{min} on your graph. *(2 marks)*

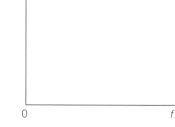

▲ **Figure 1**

3 The work function of a certain metal plate is 0.85 eV.

   a Calculate the threshold frequency of the incident radiation for this metal. *(2 marks)*

   b Calculate the maximum kinetic energy of photoelectrons emitted from this plate when light of wavelength 420 nm is directed at the metal surface. *(3 marks)*

4 Electrons are emitted from a certain metal surface when blue light is directed at its surface but not when red light is used instead. Explain why photoelectrons are emitted using blue light but not using red light. *(3 marks)*

5 A certain type of atom has excitation energies of 0.5 eV, 2.1 eV, and 4.6 eV.

   a Sketch an energy level diagram for the atom using these energy values. *(2 marks)*

   b Calculate the possible photon energies from the atom when it de-excites from the 4.6 eV level. Use a downward arrow to indicate on your diagram the energy change responsible for each photon energy. *(3 marks)*

6 Explain why the line spectrum of an element is unique to that element and can be used to identify it. *(4 marks)*

7 a State whether or not each of these experiments demonstrates the wave nature or the particle nature of matter or of light.

      i   The photoelectric effect.

      ii Electron diffraction. *(1 mark)*

   b Calculate the momentum and velocity of:

      i   an electron that has a de Broglie wavelength of 500 nm

      ii a proton that has the same de Broglie wavelength. *(4 marks)*

4.1 Waves and vibrations

Specification reference: 3.3.1.1; 3.3.1.2

Synoptic link

You have met the full spectrum of electromagnetic waves in more detail in Topic 1.3, Photons.

▲ **Figure 1** *Longitudinal waves on a slinky*

Revision tip

Vibrations of the particles in a longitudinal wave are in the same direction as that along which the wave travels.

direction of travel ⟶

hand moved from side
to side repeatedly

▲ **Figure 2** *Making rope waves*

Key term

Plane-polarised waves are transverse waves that vibrate in one plane only.

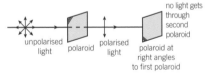

unpolarised
light   polaroid   polarised
light   polaroid at
right angles
to first polaroid
no light gets
through
second
polaroid

▲ **Figure 3** *Explaining polarisation*

Types of waves

Mechanical waves are vibrations of the particles of a substance which pass through the substance. Sound waves, seismic waves, and waves on strings are examples of mechanical waves that pass through a substance.

Electromagnetic waves are electric and magnetic waves that progress through space without the need for a substance. Electromagnetic waves include radio waves, microwaves, infrared radiation, light, ultraviolet radiation, X-rays, and gamma radiation.

Longitudinal and transverse waves

Longitudinal waves are waves in which the direction of vibration of the particles is *parallel* to (along) the direction in which the wave travels. Sound waves, primary seismic waves, and compression waves on a slinky toy are all longitudinal waves. Figure 1 shows how to send longitudinal waves along a slinky.

Transverse waves are waves in which the direction of vibration is *perpendicular* to the direction in which the wave travels. Electromagnetic waves, secondary seismic waves, and waves on a string or a wire are all transverse waves. Figure 2 shows how to send transverse waves along a rope.

Polarisation

Transverse waves are **plane-polarised** if the vibrations stay in one plane only. If the plane of the vibrations changes, the waves are **unpolarised**. Longitudinal waves cannot be polarised.

The plane of polarisation of an electromagnetic wave is defined as the plane in which the electric field oscillates. Light from a filament lamp is unpolarised and can be polarised by passing it through a polaroid filter because the filter only allows through light that vibrates in a certain direction, according to the alignment of its molecules. If the polarised light is passed through another polaroid filter, and the molecules in the two filters are aligned in perpendicular directions, the transmitted intensity is zero.

Summary questions

1 Classify the following types of waves as either longitudinal or transverse and mechanical or electromagnetic:
   a light   b infrared radiation
   c primary seismic waves   d sound waves.   (3 marks)

2 a Explain what is meant by a 'longitudinal' wave.   (1 mark)
   b Explain why a series of compressions and rarefactions travel along a slinky coil when one end of the coil is moved forwards and backwards repeatedly.   (3 marks)

3 a State two differences between a transverse wave and a longitudinal wave.   (2 marks)
   b A transverse wave travels along a horizontal rope from left to right.
      i Sketch a snapshot of the wave, indicating the direction in which the wave is travelling.
      ii Mark a point X on the crest of a wave and indicate the direction in which the rope is moving at X.   (2 marks)

4.2 Measuring waves

Specification reference: 3.3.1.1

The following terms are used to describe waves. See Figure 1 also.

- The **displacement** of a vibrating particle is its distance and direction from its equilibrium position.
- The **amplitude** of a wave is the maximum displacement of a vibrating particle.
- The **wavelength** λ of a wave is the least distance between two vibrating particles with the same displacement and velocity at the same time.
- One complete **cycle** of a wave is from maximum displacement to the next maximum displacement.
- The **frequency** f of a wave is the number of complete waves passing a point per second. The unit of frequency is the hertz (Hz).
- **wave speed** $c = f\lambda$ for waves of frequency f and wavelength λ.

The **period** of a wave is the time for one complete wave to pass a fixed point. For waves of frequency f, the period of the wave $= \dfrac{1}{f}$

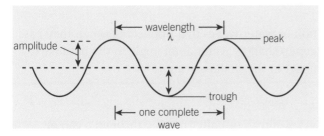

▲ **Figure 1** *Parts of a wave*

Phase difference

For two points at distance d apart along a wave of wavelength λ

$$\text{the phase difference in radians} = \frac{2\pi d}{\lambda}$$

> **Revision tip**
> The symbol c is used as a general symbol for the speed of waves. In the data booklet, c refers to the speed of electromagnetic waves in free space.

> **Revision tip**
> Amplitude is measured from the equilibrium position to maximum positive or maximum negative, not from maximum positive to maximum negative.

> **Revision tip**
> Phase difference is measured as an angle (in radians or degrees), not in terms of wavelength.
>
> Note that 1 cycle = 360° = 2π **radians**.

> **Revision tip**
> Phase difference can be stated in radians, degrees or in fractions of a cycle (e.g., a phase difference = 0.5π radians = 90° = one − quarter of a cycle).
>
> If asked for a phase difference in radians, always give your answer in radians.

Summary questions

1 a Electromagnetic waves travel through the air at a speed of $3.00 \times 10^8 \, \text{m s}^{-1}$. Calculate the frequency of light waves of wavelength 320 nm. *(1 mark)*

   b Sound waves travel through water at a speed of approximately $1500 \, \text{m s}^{-1}$. Estimate the wavelength of sound waves of frequency 20 kHz in water. *(1 mark)*

2 a State what is meant by the frequency of a wave. *(1 mark)*

   b A motor boat travelling at a constant speed creates waves that travel at a speed of $2.9 \, \text{m s}^{-1}$ and have a frequency of 1.6 Hz. Calculate the period and the wavelength of the waves. *(2 marks)*

3 Figure 2 represents a progressive wave travelling from left to right.

   a Determine the phase difference between adjacent particles in:
      i degrees **ii** radians. *(1 mark)*

   b Determine the phase difference between Q and S in radians. *(1 mark)*

   c Compare P and R in Figure 2 and half a cycle later in terms of their amplitude, phase difference, and displacement. *(3 marks)*

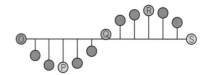

▲ **Figure 2** *Progressive waves*

4.3 Wave properties 1

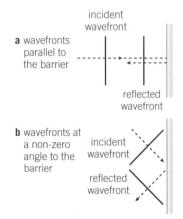

a wavefronts parallel to the barrier

b wavefronts at a non-zero angle to the barrier

▲ **Figure 1** *Reflection of plane waves*

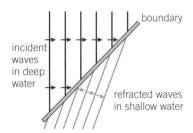

▲ **Figure 2** *Refraction*

> **Revision tip**
>
> In Figure 2, the refracted wavefront doesn't keep up with where it would have got to if its speed had not decreased at the boundary.

▲ **Figure 3** *The effect of the gap width*

> **Synoptic link**
>
> You will meet reflection, refraction, and diffraction of light in Topics 5.1, Refraction of light, 5.2, More about refraction, and 5.3, Total internal reflection.

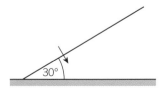

▲ **Figure 4**

Wave properties such as reflection, refraction, and diffraction occur with many different types of waves. A **ripple tank** may be used to study these wave properties. The waves observed in a ripple tank are referred to as **wavefronts**, which are lines of constant phase (e.g., crests). The direction in which a wave travels is at right angles to the wavefront.

Reflection

Straight waves directed at a certain angle to a hard flat surface (the reflector) reflect off at the same angle, as shown in Figure 2. The direction of the reflected wave is at the same angle to the reflector as the direction of the incident wave.

Refraction

When waves pass across a boundary at which the wave speed changes, the wavelength also changes. If the wavefronts approach at an angle to the boundary, they change direction as well as changing speed. This effect is known as **refraction**.

Figure 2 shows the refraction of water waves in a ripple tank when they pass across a boundary from deep to shallow water at an angle to the boundary. Note the wavelength is smaller in the shallow water.

Diffraction

Diffraction occurs when waves spread out after passing through a gap or round an obstacle. The effect can be seen in a ripple tank when straight waves are directed at a gap, as shown in Figure 3. The narrower the gap, the more the waves spread out. The longer the wavelength, the more the waves spread out.

Dish design

The bigger a satellite dish is, the stronger the signal it can receive, because more radio waves are reflected by the dish onto the aerial. But a bigger dish diffracts the waves less so it needs to be aligned more carefully than a smaller dish, otherwise it will not focus the radio waves onto the aerial.

> **Summary questions**
>
> 1 A straight wave in a ripple tank is directed at an angle of 30° to a flat reflector, as shown in Figure 4.
> Copy the diagram accurately and show the position and direction of the wavefront after it reflects from the reflector. (*1 mark*)
>
> 2 A straight wave is directed at an angle of 30° to a straight boundary where the wave speed increases.
> a Draw the straight wave and the boundary before the wave crosses the boundary. (*1 mark*)
> b Draw the straight wave when half of it has travelled across the boundary. Indicate on your diagram the direction of the part of the wavefront that has crossed the boundary. (*1 mark*)

4.4 Wave properties 2
Specification reference: 3.3.1.1; 3.3.1.2

The principle of superposition

When waves meet, they pass through each other. Where they meet, they combine for an instant before moving apart. This effect is called **superposition**.

The principle of superposition states that when two waves meet, the total displacement at a point is equal to the sum of the individual displacements at that point.

Stationary waves on a rope

Stationary waves are formed on a rope if two people send waves continuously along a rope from either end, as shown in Figure 1. The two sets of waves are referred to as **progressive waves** to distinguish them from stationary waves. They combine at fixed points along the rope to form points of no displacement or **nodes** along the rope. At each node, the two sets of waves are always 180° out of phase, so they cancel each other out.

Water waves in a ripple tank

Figure 2 shows two sets of circular waves produced in a ripple tank.

- Points of cancellation are created where a crest from one dipper meets a trough from the other dipper. At each point, the waves from one dipper are 180° out of phase with the waves from the other dipper.

- Points of reinforcement are created where a crest meets a crest, or where a trough meets a trough. At each point, the waves from one dipper are in phase with the waves from the other dipper.

Provided the two sets of waves have the same frequency and have a constant phase difference, cancellation and reinforcement occurs at fixed positions. This effect is known as **interference**. The dippers act as **coherent sources** because they vibrate at the same frequency with a **constant phase difference**.

▲ **Figure 1** *Making stationary waves*

▲ **Figure 2** *Interference of water waves*

Common misconception

A common mistake is to say coherent sources must have a phase difference of zero. This is wrong – the phase difference must be constant!

Key term

Superposition is the combining of waves when they meet for an instant before they move apart.

Summary questions

1 a State the principle of superposition of waves. *(1 mark)*
  b Describe the difference between a progressive wave and a stationary wave on a rope. *(1 mark)*

2 State and explain how you would expect the interference pattern in Figure 2 to change if the frequency of the waves produced by the dippers is increased? *(2 marks)*

3 Microwaves from a transmitter are directed at two parallel slits in a metal plate (see Figure 3). The two slits act as coherent sources of microwaves.
  a State what is meant by coherent sources. *(1 mark)*
  b A detector is placed on the other side of the metal plate on a line XY parallel to the plate. When the detector is moved steadily along the line, the detector signal decreases then increases again several times. Explain why the signal changes in this way. *(3 marks)*

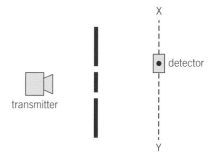

▲ **Figure 3**

4.5 Stationary and progressive waves

Specification reference: 3.3.1.3

Synoptic link

Stationary waves on a string are discussed in more detail in Topic 4.6 , More about stationary waves on strings.

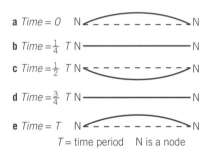

a *Time = 0*   N━━━━━━━N
b *Time =$\frac{1}{4}$ T*   N━━━━━━━N
c *Time =$\frac{1}{2}$ T*   N━━━━━━━N
d *Time =$\frac{3}{4}$ T*   N━━━━━━━N
e *Time = T*   N━━━━━━━N

T = time period   N is a node

▲ **Figure 1** *First harmonic vibrations*

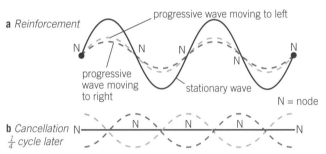

a *Reinforcement*

progressive wave moving to left

progressive wave moving to right

stationary wave

N = node

b *Cancellation*
$\frac{1}{4}$ *cycle later*

▲ **Figure 2** *Explaining stationary waves*

Revision tip

In general, in any stationary wave pattern:

1 The amplitude of a vibrating particle varies with position from zero at a node to maximum amplitude at an antinode.

2 The phase difference between two vibrating particles is:

- zero if the two particles are between adjacent nodes or separated by an even number of nodes

- 180° (= π radians) if the two particles are separated by an odd number of nodes.

3 The distance between adjacent nodes =$\frac{1}{2}\lambda$ where λ is the wavelength of the progressive waves that form the nodes.

Formation of stationary waves

A stationary wave is formed when two progressive waves of the same frequency pass through each other. The stationary wave formed has equally-spaced **nodes** (points of no displacement) and **antinodes** (points of maximum displacement) at fixed positions.

1 Stationary waves can be formed on a string in tension by fixing both ends and making the string vibrate. As a result, progressive waves travel towards each end, reflect at the ends, and then pass through each other to form a stationary wave pattern on the string. Different stationary wave patterns occur if the string is made to vibrate at different frequencies.

2 Sound resonates at certain frequencies in an air-filled tube or pipe. The sound waves in the pipe reflect at the ends and the reflected waves pass through each other. In a pipe closed at one end, resonance occurs if there is an antinode at the open end and a node at the closed end.

3 Microwaves form a stationary wave pattern when they are reflected by a metal plate back towards the transmitter. When a detector is moved between the transmitter and the metal plate, the detector signal is least at the nodes.

Explanation of stationary waves

Figure 2 shows two progressive waves passing through each other.

- When the two waves reinforce each other, they produce a large wave.

- A quarter of a cycle later, the two waves have each moved one-quarter of a wavelength in opposite directions. They are now in antiphase so the two waves cancel each other.

- After a further quarter cycle, the two waves reinforce each other again.

Note that the position of the points of no displacement, the nodes, does not change.

Summary questions

1 a State the conditions necessary for the formation of a stationary wave. *(1 mark)*

  b The stationary wave pattern in Figure 1 was set up on a string of length 0.60 m when it was made to vibrate at a frequency of 150 Hz.
  Calculate **i** the wavelength **ii** the speed of the progressive waves on the string. *(1 mark each)*

2 Describe the differences between a stationary wave and a progressive wave at different positions along each wave in terms of:
  a the amplitude of the wave *(2 marks)*
  b the phase difference. *(2 marks)*

3 A small loudspeaker connected to a signal generator was used to send sound waves into an air-filled glass tube which was closed at its other end. Explain why the tube resonated with sound at certain frequencies. *(3 marks)*

4.6 More about stationary waves on strings

Specification reference: 3.3.1.3

Stationary waves on a vibrating string

Figure 1 shows how stationary waves can be set up on a string in tension. As the frequency of the generator is increased from a very low value, different stationary wave patterns are seen on the string. In every case, the string has a node at either end.

The first harmonic pattern of vibration is the lowest possible frequency that gives the pattern shown in Figure 1a. The distance between adjacent nodes is $\frac{1}{2}\lambda_1$, so for a string of length L:

- the first harmonic wavelength $\lambda_1 = 2L$
- the first harmonic frequency $f_1 = \frac{c}{\lambda_1} = \frac{c}{2L}$, where c is the speed of the progressive waves on the wire.

The mth harmonic (where m is a whole number) has m equal loops along its length which means:

- the mth harmonic wavelength $\lambda_m = \frac{2L}{m}$ because each loop has a length of half a wavelength
- the mth harmonic frequency $f_m = \frac{c}{\lambda_m} = \frac{mc}{2L} = mf_1$.

In general, stationary wave patterns occur at frequencies f_1, $2f_1$, $3f_1$, $4f_1$, and so on, where f_1 is the first harmonic frequency of the first harmonic vibrations.

The frequency of the first harmonic of a vibrating string or wire is given by

$$f_1 = \frac{1}{2L}\sqrt{\frac{T}{\mu}}$$

where T is the tension in the wire and μ is its mass per unit length

string at maximum displacement

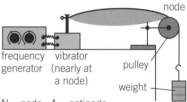

node

frequency generator
vibrator (nearly at a node)
pulley
weight

N = node   A = antinode
(dotted line shows string half a cycle earlier)

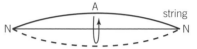

a *First harmonic*

b *Second harmonic*

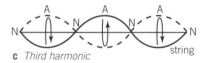

c *Third harmonic*

▲ **Figure 1** *Stationary waves on a string*

Maths skill

For a uniform wire of length L and area of cross-section A, its volume $V = LA$.

So its mass m = density ρ × volume $V = \rho LA$.

Therefore, its mass per unit length $\mu = \frac{m}{L} = \rho A$.

Summary questions

1. A stretched wire of length 0.760 m at a tension of 64 N vibrates at its first harmonic with a frequency of 280 Hz. Calculate:
   a. the wavelength of the waves on the wire *(1 mark)*
   b. the mass per unit length of the wire. *(2 marks)*

2. For the wire in **Q1** at the same tension, calculate the length of the wire to produce a frequency of 384 Hz. *(3 marks)*

3. Two stretched wires X and Y that have the same length are under the same tension. The diameter of X is half that of Y and the density of X is 8 times that of Y. The frequency of the 1st harmonic of X is 100 Hz. Calculate the frequency of the 1st harmonic of Y. *(3 marks)*

Synoptic link

See Topic 11.1, Density, for more about density

Revision tip

The data is given to 2 significant figures so the final answer is given to 2 significant figures. To avoid a 'rounding' error in the final answer, data from each step of the calculation should be carried forward with one more significant figure and only rounded off in the final answer.

4.7 Using an oscilloscope

Using an oscilloscope

Although the use of an oscilloscope is in the full A Level specification and is not part of the AS specification, it is included in this Revision Guide and in the Student Book to give you information about how to measure sound waves and determine their frequency and amplitude (i.e., the peak value).

Displaying a waveform

- The oscilloscope's **time base circuit** is switched on to make the spot move at constant speed left to right across the screen, then back again much faster. Because the spot moves at constant speed across the screen, the x-scale is calibrated, usually in milliseconds or microseconds per centimetre.

- The pd to be displayed is connected to the Y-plates via the Y-input so the spot moves up and down as it moves left to right across the screen. As it does so, it traces out the waveform on the screen. Because the vertical displacement of the spot is proportional to the pd applied to the Y-plates, the Y-input is calibrated in volts per centimetre (or per division if the grid on the oscilloscope screen is not a centimetre grid). The calibration value is usually referred to as the Y-sensitivity or Y-gain of the oscilloscope.

Figure 1 shows the trace produced when an alternating pd is applied to the Y-input. The screen is marked with a centimetre grid.

To measure the peak pd, observe that the waveform height from the bottom to the top of the wave is 3.2 cm. The amplitude (i.e., peak height) of the wave is therefore 1.6 cm. As the Y-gain is set at $5.0\,\mathrm{V\,cm^{-1}}$, the peak pd is therefore $8.0\,\mathrm{V}\ (= 5.0\,\mathrm{V\,cm^{-1}} \times 1.6\,\mathrm{cm})$.

We can see from the waveform that one full cycle corresponds to a distance of 3.8 cm across the screen horizontally. As the time base control is set at $2\,\mathrm{ms\,cm^{-1}}$, the time period T is therefore $7.6\,\mathrm{ms}\ (= 2\,\mathrm{ms\,cm^{-1}} \times 3.8\,\mathrm{cm})$. Therefore, the frequency of the alternating pd is 132 Hz.

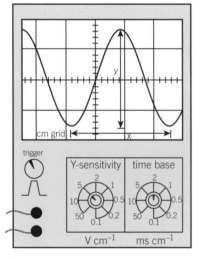

▲ **Figure 1** *Using an oscilloscope*

Summary questions

1 An alternating pd is applied to the Y-input of an oscilloscope. The height of the waveform from the bottom to the top is 5.2 cm when the Y-gain is $0.2\,\mathrm{V\,cm^{-1}}$. Calculate the peak value of the alternating pd. *(1 mark)*

2 The time base control of an oscilloscope is set at $5.0\,\mathrm{ms\,cm^{-1}}$ and an alternating pd is applied to the Y-input. The horizontal distance across four complete cycles is observed to be 6.4 cm. Calculate the frequency of the alternating pd. *(2 marks)*

1   **a**   Explain why a transverse wave can be polarised whereas a longitudinal wave cannot be polarised.   (*2 marks*)

     b   A filament lamp is observed through two pieces of polaroid which are initially aligned parallel to each other. Describe and explain what you would expect to observe when one of the polaroids is rotated through 360°.   (*3 marks*)

2   A microwave transmitter directs waves towards a metal plate, as shown in Figure 1. When a microwave detector is moved along a line normal to the transmitter and the plate, the detector signal increases and decreases as it moves through a sequence of equally spaced maxima and minima.

metal plate

transmitter   detector

▲ **Figure 1**

     a   Explain why the detector signal varies in this way as the detector is moved along the line.   (*4 marks*)

     b   The detector is placed at a position where the intensity is a minimum. When it is moved through a distance of 82 mm, it passes through four other minima and reaches a further minimum.

         Calculate the wavelength of the microwaves.   (*3 marks*)

3   A stretched string of length L fixed at both ends is made to vibrate so that it forms a stationary wave consisting of four equal loops.

     a   **i**   Sketch the pattern of vibration of the string and mark the position of the nodes and antinodes on the string.

        ii   Compare the amplitude and state the phase difference of the vibrations of points X and Y on the string at distances $\frac{L}{3}$ from each ends.   (*4 marks*)

     b   The length L of the string is 800 mm and its mass per unit length is $7.4 \times 10^{-4}\,\mathrm{kg\,m^{-1}}$.

        i   Calculate the wavelength of the progressive waves on the string.

        ii   The string vibrates at a frequency of 200 Hz. Calculate the tension in the string.   (*6 marks*)

4   An ultrasound probe in an ultrasound scanner emits ultrasound waves of frequency 1.6 MHz

     a   The speed of ultrasound in body tissue is approximately $1500\,\mathrm{m\,s^{-1}}$. Estimate the wavelength of the ultrasound waves in the body.   (*1 mark*)

     b   The probe sends ultrasound pulses of duration 10 μs into the body.

        i   Calculate the number of ultrasound waves in each pulse.

        ii   The pulses from the probe are reflected back to the probe by tissue boundaries in the body. Give *two* reasons why the amplitude of each pulse that returns to the probe is smaller than its amplitude when it entered the body.   (*4 marks*)

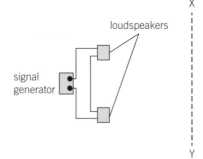

loudspeakers

X

signal generator

Y

▲ **Figure 2**

string at maximum displacement

node

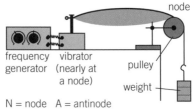

frequency generator

vibrator (nearly at a node)

pulley

weight

N = node   A = antinode
(dotted line shows string half a cycle earlier)

▲ **Figure 3**

5 Two small loudspeakers connected to a signal generator act as coherent sources of sound waves, as shown in Figure 2.

  a Explain what is meant by coherent sources. (*1 mark*)

  b A student walks along the straight line XY in front of the loudspeakers and notices the sound intensity due to the loudspeakers is a minimum at regular intervals along the line. Explain why the sound intensity is a minimum at these positions. (*2 marks*)

6 A small loudspeaker connected to a signal generator is placed near the open end of a pipe of length 380 mm, which is closed at its other end. When the output frequency of the signal generator is increased from zero, the pipe resonates with sound at different frequencies.

  a i The minimum frequency at which the pipe resonates with sound is 225 Hz. At this frequency, a stationary wave is set up in the pipe with one node only which is at the closed end and an antinode very close to the open end. Calculate the wavelength of the sound in the pipe and determine the speed of sound in the pipe.

  ii Describe how the amplitude of vibration of the air particles in the pipe changes along the pipe. (*3 marks*)

  b Explain why the pipe resonates with sound at a frequency of about 675 Hz. (*2 marks*)

7 Figure 3 shows an arrangement used to set up stationary waves on a stretched string.

  a Describe how you would use this arrangement to investigate how the frequency of vibration of the first harmonic, f_1, of a stretched string varies with the tension T in the string. (*4 marks*)

  b The following results were obtained in the investigation in part **a** using a string of length 0.64 m.

$T/$ N	1.0	1.5	2.0	2.5	3.0	3.5	4.0
$f_1/$ Hz	42	52	60	67	73	79	84

  i The first harmonic frequency is given by the equation

$$f_1 = \frac{1}{2L}\sqrt{\frac{T}{\mu}}$$

  Use the equation to show that a graph of f_1^2 against T should be a straight line through the origin. (*2 marks*)

  ii Plot a graph of f_1^2 against T to determine μ, the mass per unit length of the string. (*6 marks*)

5.1 Refraction of light
5.2 More about refraction

Specification reference: 3.3.2.3

Refraction is the change of direction that occurs when light passes non-normally across a boundary between two transparent substances. Figure 1 shows the change of direction of a light ray when it enters and when it leaves a rectangular glass block in air. For a light ray travelling from air into a transparent substance

$$\text{the refractive index of the substance, } n = \frac{\sin i}{\sin r}$$

Notice that **partial reflection** can also occur when a light ray is refracted.

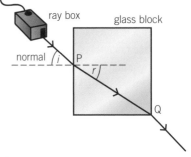

▲ **Figure 1** Investigating refraction

A general rule for refraction

Figure 2 shows a light ray that passes across a straight boundary from a transparent substance of refractive index n_1 into another transparent substance of refractive index n_2. If θ_1 and θ_2 are the angles of incidence and refraction respectively, it can be shown that $n_1 \sin \theta_1 = n_2 \sin \theta_2$

Explaining refraction

Refraction occurs because the speed of the light waves is different in each substance. For light waves entering a transparent substance from air at a plane boundary, it can be shown that

$$\frac{\sin i}{\sin r} = \frac{c}{c_s}$$

where c is the speed of the waves in a vacuum and c_s is the speed in the substance.

Therefore, the refractive index of the substance

$$n_s = \frac{c}{c_s}$$

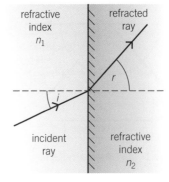

▲ **Figure 2** The n sin θ rule

Summary questions

1   A glass block has a refractive index of 1.52. A light ray enters the block from air at point X at an angle of incidence of 30.0°. Calculate the angle of refraction of the light ray.   (*2 marks*)
2   The light ray in **Q1** is refracted again at point Y where it leaves the glass block and enters the air. The angle of refraction of the light ray at Y is 70.0°. Calculate the angle of incidence of the light ray at Y.   (*2 marks*)

3   A light ray in air was directed into a semicircular glass block at the midpoint C of the flat side at an angle, as shown in Figure 3. The refractive index of the glass block was 1.54.
    a   Explain why the light ray was not refracted when it entered the block.   (*1 mark*)
    b   The angle of incidence of the light ray at C was 35°. Calculate the angle through which the light ray was deflected at C.   (*3 marks*)

4   Calculate the angle of refraction for a light ray entering glass of refractive index 1.50 at an angle of incidence of 40.0° from water of refractive index 1.33.   (*2 marks*)

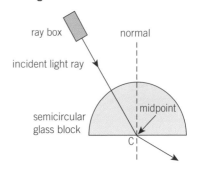

▲ **Figure 3** Explaining refraction

5.3 Total internal reflection

Specification reference: 3.3.2.3

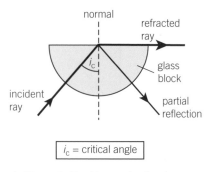

i_c = critical angle

▲ **Figure 1** *Total internal reflection*

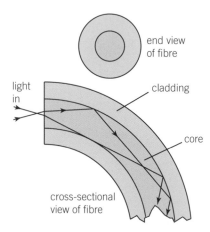

▲ **Figure 2** *Fibre optics*

When a light ray travels across a boundary between two transparent substances or between air and a transparent substance, it undergoes **total internal reflection** at the boundary if the incident substance has a larger refractive index than the other substance and the angle of incidence exceeds the critical angle.

At the critical angle i_c, the angle of refraction is 90°. Therefore, $n_1 \sin i_c = n_2 \sin 90$ where n_1 is the refractive index of the incident substance and n_2 is the refractive index of the other substance. Since sin 90 = 1, then

$$\sin i_c = \frac{n_2}{n_1}$$

Optical fibres

Figure 2 shows the path of a light ray along an optical fibre. The light ray is totally internally reflected each time it reaches the fibre boundary.

1 A **communications optical fibre** transmits pulses of light from a transmitter at one end to a receiver at the other end. The material needs to be highly transparent to minimise absorption which would reduce the amplitude of light pulses travelling along the fibre. Each fibre consists of a core surrounded by cladding of lower refractive index. Total internal reflection takes place at the core–cladding boundary.

   • Without cladding, light would cross between fibres where they touch. This would reduce the amplitude of the pulses and the signals would not be secure.

   • Monochromatic light and fibres with a very narrow core are used to prevent pulse lengthening which would cause pulses to merge. **Modal** (i.e., multipath) **dispersion** is reduced by the very narrow core. In a wide core, non-axial rays undergo more total internal reflections than axial rays and therefore take longer, making the pulses longer. **Material** (or spectral) **dispersion** occurs if white light is used. Violet light is slower than red light in glass. The violet component of a white light pulse falls behind the red component so the pulse lengthens.

2 The **medical endoscope** contains two bundles of fibres, one of which is a **coherent bundle**, which means the fibre ends are in the same relative positions at each end. Light is sent through the non-coherent bundle into the body cavity to be observed. A lens over the end of the coherent bundle forms an image of the body cavity on the end of the fibre bundle. The image is observed at the other end of the coherent fibre bundle.

Summary questions

1 Calculate the critical angle for glass of refractive index 1.54:
   a and air **b** and water (refractive index = 1.33). (*2 marks each*)

2 **a** Explain why optical fibres used for communications have a very narrow core surrounded by less refractive cladding. (*3 marks*)
   b An optical fibre has a core of refractive index 1.51 surrounded by cladding of refractive index 1.42. Calculate the critical angle of the core–cladding boundary. (*2 marks*)

3 **a** Explain what is meant by a coherent bundle in a medical endoscope. (*1 mark*)
   b Explain why a medical endoscope has two fibre bundles. (*2 marks*)

5.4 Double slit interference

Specification reference: 3.3.2.1

Interference of light can be observed by directing light from a suitable light source at two closely spaced parallel slits (double slits).

- The two slits act as **coherent** sources of waves, which means that they emit light waves with a constant phase difference and the same frequency.
- Alternate bright and dark fringes, referred to as **Young's fringes**, can be seen on a white screen placed where the diffracted light from the double slits overlaps. The fringes are evenly spaced and parallel to the double slits.

Where a *bright* fringe is formed, the light from one slit reinforces the light from the other slit. In other words:

- the light waves from each slit arrive *in phase* with each other
- the path difference = $m\lambda$ where $m = 0, 1, 2$, and so on.

Where a *dark* fringe is formed, the light from one slit cancels the light from the other slit. In other words:

- the light waves from the two slits arrive *180° out of phase*
- the path difference = $\left(m + \dfrac{1}{2}\right)\lambda$.

Warning: Never look along a laser beam, even after reflection. If a laser is used as the light source, the fringes must be displayed on a screen because a beam of laser light will damage the retina if it enters the eye.

The double slit equation

$$\textbf{fringe separation } w = \frac{\lambda D}{s}$$

where λ is the wavelength of light, s is the slit spacing, and D is the distance from the slits to the screen.

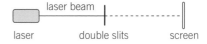

▲ **Figure 1** *Using a laser to demonstrate interference*

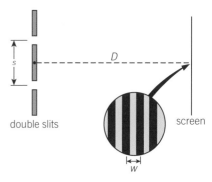

▲ **Figure 2** *Diagram to show w, D, and s for a Young's double slit experiment*

Summary questions

1. In a double slit experiment using red light, a fringe pattern is observed on a screen at a fixed distance from the double slits. How would the fringe pattern change if:
   a. the screen is moved further from the slits *(1 mark)*
   b. blue light is used instead? *(1 mark)*

2. The following measurements were made in a double slit experiment: centre-to-centre distance across 5 dark fringes = 4.5 mm, slit spacing $s = 0.40$ mm, slit–screen distance $D = 0.85$ m. Calculate the wavelength of light used. *(3 marks)*
   Hint: Remember that the number of fringe spacings ≠ the number of fringes.

3. In **Q2**, state and explain two differences in the appearance of the fringes if light of a longer wavelength was used. *(2 marks)*

5.5 More about interference

Coherence

Interference fringes are produced using a double slit (i.e., two narrow slits close together), provided the slits emit wavefronts with a constant phase difference. The slits must act as **coherent sources** otherwise the points of cancellation and reinforcement change at random.

- A laser is a source of coherent light. So the two slits act as coherent sources when a laser beam is directed at them.
- Light from a single narrow slit illuminated by a filament lamp can be used to produce interference fringes instead of laser light. Each wave crest or wave trough from the single slit always passes through one of the double slits a fixed time after it passes through the other slit. So the waves from the double slits always have a constant phase difference.
- Light from two nearby lamp bulbs could not form an interference pattern because a filament lamp emits light waves at random. The points of cancellation and reinforcement would change at random, so no interference pattern is possible.

Wavelength and colour

In the double slit experiment, the fringe separation depends on the colour of light used. Each colour of light has its own narrow band of wavelengths. The fringe separation is greater for red light than for blue light because red light has a longer wavelength than blue light.

White light fringes

For white light, each component colour of white light produces its own fringe pattern, and each pattern is centred on the screen at the same position. As a result:

- the central bright fringe is white because every colour contributes at the centre of the pattern
- the inner bright fringes are tinged with blue on the inner side and red on the outer side. This is because the red fringes are more spaced out than the blue fringes so the two fringe patterns do not overlap exactly
- the outer bright fringes become fainter with increasing distance from the centre. Also, the fringes merge because the colours overlap.

> ### Key term
>
> **Coherent sources** emit light waves of the same frequency with a constant phase difference.

Summary questions

1 a Explain what is meant by 'coherent light sources'. *(1 mark)*

   b Explain why an interference pattern cannot be seen using two filament lamps side by side. *(2 marks)*

2 Double slit interference fringes are observed using light of wavelength 635 nm and a double slit of slit spacing 0.45 mm. The fringes are observed on a screen at a distance of 1.90 m from the double slits. Calculate the fringe separation of these fringes. *(2 marks)*

3 A double slit is used to observe the interference fringes from a 12 V filament lamp connected to a variable 0–12 V voltage supply. When the voltage across the filament lamp is increased, the colour of the filament changes from red to white. Describe how the fringe pattern alters in this change. *(3 marks)*

5.6 Diffraction

Specification reference: 3.3.2.2

Observing diffraction

Diffraction is the spreading of waves when they pass through a gap or by an edge. When waves pass through a gap:

1   The diffracted waves spread out more if the gap is made narrower or the wavelength is made larger.

2   Each diffracted wavefront has breaks either side of the centre. These breaks are due to waves diffracted by adjacent sections on the gap being out of phase and cancelling each other out in certain directions.

Diffraction is a general property of all waves and is very important in the design of optical instruments such as cameras, microscopes, and telescopes. This is because less diffraction occurs when waves pass through a wide gap than through a narrow gap. Therefore, more detail can be seen in optical images formed by wide lenses.

> **Synoptic link**
>
> You have met the study of water waves using a ripple tank in more detail in Topic 4.3, Wave properties 1.

Diffraction of light by a single slit

When a parallel beam of light is directed at a single slit, the diffracted light forms a pattern that can be observed on a white screen, as shown in Figure 1. The pattern shows a central fringe with further fringes either side of the central fringe.

*   The central fringe is twice as wide as each of the outer fringes (measured from minimum to minimum intensity).
*   Each of the outer fringes is the same width.
*   The fringes decreases in intensity with distance from the centre.
*   The central fringe has a much higher peak intensity than the other fringes.
*   The width of each fringe is proportional to $\frac{\lambda}{a}$, where a is the width of the slit.

Single slit diffraction and Young's fringes

In general, for monochromatic light of wavelength λ incident on two slits of aperture width a at slit separation s (from centre to centre), the intensity peaks of the interference fringes are affected by single slit diffraction as shown in Figure 2. If the aperture width a is made narrower, more diffraction occurs so more interference fringes are seen between adjacent single slit minima as the single slit minima will be further apart.

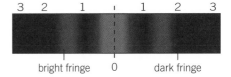
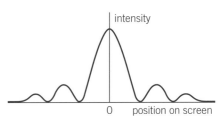

▲ **Figure 1** *Single slit diffraction*

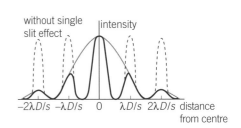

▲ **Figure 2** *Intensity distribution for Young's fringes*

> **Summary questions**
>
> 1   A narrow beam of monochromatic light is directed normally at a single slit. The diffracted light forms a fringe pattern on a white screen.
>   a   Describe the appearance of the fringe pattern on the screen. *(3 marks)*
>   b   Sketch a graph to show how the intensity of the fringes varies with distance across the screen. *(2 marks)*
>
> 2   Single slit diffraction fringes are observed using red light. State and explain how the pattern changes if:
>   a   the slit is made narrower *(2 marks)*
>   b   blue light is used instead of red light. *(2 marks)*

5.7 The diffraction grating

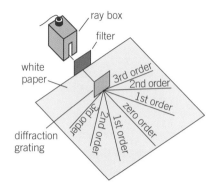

▲ **Figure 1** *The diffraction grating*

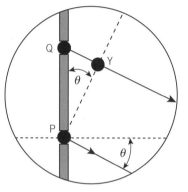

▲ **Figure 2** *The nth order wavefront*

Worked example

When monochromatic light is directed normally at a diffraction grating which has 600 lines per millimetre, a second order beam is observed at an angle of diffraction of 45.0°. Calculate the wavelength of the light.

Solution

The grating has 600 000 lines per metre. Hence $d = \dfrac{1}{600\,000}$ m

$= 1.67 \times 10^{-6}$ m

Therefore $2\lambda = d\sin\theta = 1.67 \times 10^{-6} \times \sin 45.0 = 1.18 \times 10^{-6}$ which gives $\lambda = 5.90 \times 10^{-7}$ m

Synoptic link

You have met line spectra in more detail in Topic 3.5, Energy levels and spectra.

Revision tip

Reminder: 1 degree = 60 minutes.

Testing a diffraction grating

A **diffraction grating** consists of a plate with many closely spaced parallel slits ruled on it. When a parallel beam of monochromatic light is directed normally at a diffraction grating, light is transmitted by the grating in certain directions only. This is because: the light passing through each slit is diffracted and the diffracted light waves from adjacent slits reinforce each other in certain directions only.

The diffraction grating equation

For light of wavelength λ incident normally on a grating which has N slits per metre, the angle of diffraction θ of the nth order beam is given by the equation

$$d\sin\theta = n\lambda$$

where the grating spacing $d = \dfrac{1}{N}$

Note: The maximum order number is given by the value of $\dfrac{d}{\lambda}$ rounded down to the nearest whole number.

Proof of the diffraction grating equation

Look at Figure 2. The wavefront emerging from slit P reinforces a wavefront emitted n cycles earlier by the adjacent slit Q. The distance QY is therefore equal to $n\lambda$.

The angle of diffraction of the beam θ is equal to the angle between the wavefront and the plane of the slits so QY = QP $\sin\theta$. Since QY = $n\lambda$ and QP = d (the grating spacing), it follows that $n\lambda = d\sin\theta$.

Types of spectra

1 Continuous spectra

The spectrum of light from a filament lamp is a continuous spectrum of colour from deep violet at about 350 nm to deep red at about 650 nm.

2 Line emission spectra

A glowing gas emits light that has a spectrum consisting of narrow vertical lines of different colours. The wavelengths of the lines are characteristic of the chemical element that produced the light.

3 Line absorption spectra

A line absorption spectrum is a continuous spectrum with narrow dark lines at certain wavelengths. When white light passes through a glowing gas, the gas atoms absorb light of the same wavelengths at which they can emit so the transmitted light is missing these wavelengths.

Summary questions

1 A laser beam of wavelength 630 nm is directed normally at a diffraction grating with 600 lines per millimetre. Calculate:
   a the angle of diffraction of the first order beam (3 marks)
   b the maximum order number. (2 marks)

2 Light of wavelength 430 nm is directed normally at a diffraction grating. The angle of diffraction of the 1st order beam was at 28°. Calculate the angle of diffraction of the most diffracted order. (3 marks)

Chapter 5 Practice questions

1 A light ray enters an equilateral glass prism of refractive index 1.55 at the midpoint of one side of the prism at an angle of incidence of 35.0°.

   a **i**   Sketch this arrangement and show the path of the light ray in and out of the prism.

       ii Calculate the angle of refraction of the light ray in the glass.   *(3 marks)*

   b **i**   Show that the angle of incidence where the light ray leaves the glass prism is about 38°.

       ii Calculate the angle of refraction of the light ray where it leaves the prism.   *(3 marks)*

2 **a** Explain why the core of an optical fibre used in communications needs to be very narrow.   *(2 marks)*

   b **i**   The core of an optical fibre has a refractive index of 1.55. The core is surrounded by cladding of refractive index 1.45. Calculate the critical angle at the core–cladding boundary.

       ii State one advantage of cladding an optical fibre.   *(4 marks)*

3 A white light ray is directed through the curved side of a semicircular glass block at the midpoint of the flat side, as shown in Figure 1. The angle of incidence of the light ray at the flat side is 40°. The refractive index of the glass for red light is 1.52 and for blue light is 1.55.

   a Calculate the angle of refraction at the midpoint of:

      i   the red component of the light ray

      ii the blue component of the light ray.   *(4 marks)*

   b Show the angle between the red and blue components of the refracted light ray is about 7°.   *(1 mark)*

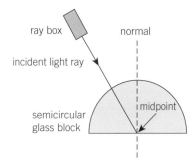

▲ Figure 1

4 A laser beam of wavelength 630 nm is directed normally at a double slit to produce an interference pattern of alternate bright and dark fringes on a screen.

   a Explain why alternate bright and dark fringes were seen on the screen.   *(4 marks)*

   b The distance across 6 bright fringes was measured to be 65 mm. The screen was at a distance of 2.50 m from the double slit. Calculate the slit spacing.   *(3 marks)*

5 Red light from a laser is directed normally at a slit that can be adjusted in width. The diffracted light from the slit forms a pattern of diffraction fringes on a white screen.

   a Sketch a graph to show how the intensity of the fringes varies across the pattern.   *(3 marks)*

   b Describe how the appearance of the fringes changes if the slit is made wider.   *(2 marks)*

6 Light directed normally at a diffraction grating has a wavelength range from 430 to 445 nm. The grating has 600 lines per mm.

   a How many diffracted orders are observed in the transmitted light?   *(4 marks)*

   b For the highest order, calculate the angular width of the diffracted beam.   *(3 marks)*

7 Young's fringes are produced on the screen using the arrangement shown in Figure 2.

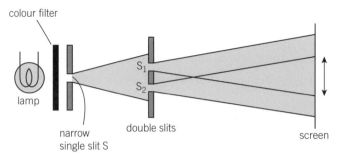

colour filter

lamp

narrow
single slit S

double slits

S_1
S_2

screen

view from above

▲ **Figure 2**

  a **i** Explain why slit S should be narrow.

    ii Explain why slits S_1 and S_2 are coherent sources? (*6 marks*)

  b The fringes are equally spaced but some bright fringes are fainter than bright fringes further from the central fringe. Explain why some bright fringes are fainter than bright fringes further from the centre. (*3 marks*)

8 A narrow beam of monochromatic light is directed normally at a diffraction grating which has 600 lines per millimetre.

  a The angle of diffraction of the 2nd order beam is 40.0°. Calculate the wavelength of the light. (*3 marks*)

  b A semicircular glass block of refractive index 1.50 is placed with its flat side against the grating on the opposite side to the incident light. The block is positioned so that the direction of the zero order beam is unchanged as shown in Figure 3. Show that the angle of diffraction of the second order beam is now about 25°. (*4 marks*)

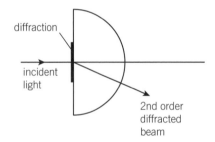

diffraction

incident
light

2nd order
diffracted
beam

▲ **Figure 3**

6.1 Vectors and scalars

Specification reference: 3.4.1.1

Representing a vector

A vector is any physical quantity that has a direction as well as a magnitude. Examples of vectors include displacement, velocity, acceleration, force, and momentum.

A scalar is any physical quantity that is not directional. Examples of scalars include distance, mass, density, volume, and energy.

Any vector can be represented as an arrow of length proportional to the magnitude of the vector quantity in the direction of the vector.

Addition of vectors

Any vectors of the same type can be added together by drawing a scale diagram or using a calculator.

1 Drawing a scale diagram

Figure 1 shows two displacement vectors OA and AB added together to give the resultant displacement vector OB.

$$\mathbf{OB} = \mathbf{OA} + \mathbf{AB}$$

2 Using a calculator

To add two perpendicular vectors F_1 and F_2:

- the magnitude of the resultant force $F = \sqrt{F_1^2 + F_2^2}$
- the angle θ between the resultant and F_1 is given by $\tan\theta = \dfrac{F_2}{F_1}$.

Resolving a vector

Any vector can be resolved into two perpendicular components. For example, Figure 2 shows a force F resolved into components along perpendicular axes, x and y, where the direction of the force is at an angle to the x-axis.

- $F_x = F\cos\theta$ **parallel to the x-axis.**
- $F_y = F\sin\theta$ **perpendicular to x-axis.**

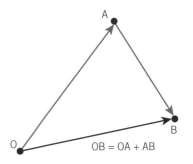

▲ **Figure 1** *Drawing a scale diagram*

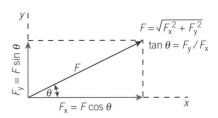

▲ **Figure 2** *Resolving a force*

Summary questions

1 An airplane takes off from airport A and lands at airport B, which is 180 km 30° due east of north from A.
   a Sketch the displacement vector AB on two perpendicular axes representing due north and due east with A at the intersection of the two axes. *(1 mark)*
   b Calculate the component of vector AB **i** due east **ii** due north. *(1 mark)*

2 Calculate the magnitude and direction of the resultant force on an object which is acted on by a force of 5.0 N and a force of 11.0 N that are:
   a in opposite directions *(1 mark)*
   b at right angles to each other. *(2 marks)*

3 A descending parachutist of total weight 900 N is acted on by a horizontal force of 150 N due to the wind. Calculate the magnitude and direction of the resultant of these two forces. *(3 marks)*

6.2 Balanced forces

Specification reference: 3.4.1.1

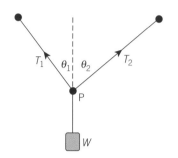

▲ **Figure 1** *A suspended weight*

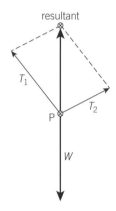

▲ **Figure 2** *The parallelogram of forces rule*

Synoptic link

The three force vectors T_1, T_2, and W in Figure 2 form a triangle. See Topic 6.6, Equilibrium rules, for more about the **triangle of forces**.

Any object at rest or moving at constant velocity is in **equilibrium**.

If *two forces only* are acting on the object, the two forces must be equal and opposite to each other for the object to be in equilibrium. The resultant of the two forces is zero and the two forces are said to be *balanced*.

If *two or more forces* are acting on the object, their combined effect (the resultant) must be zero for it to be in equilibrium. To check the resultant is zero, either:

- resolve each force along the same parallel and perpendicular lines and balance the components along each line (see the Worked example below), or

- use the parallelogram of forces rule if there are only three forces.

Worked example

An object of weight W is tied to a vertical string, which is supported by two strings at angles θ_1 and θ_2 to the vertical, as shown in Figure 1. Suppose the tension in the string at angle θ_1 to the vertical is T_1 and the tension in the other string is T_2. Derive two equations relating T_1, T_2, and W.

Solution

At the point P where the strings meet, the forces T_1, T_2, and W are in equilibrium.

Resolving T_1 and T_2 vertically and horizontally gives:

1 horizontally: $T_1 \sin \theta_1 = T_2 \sin \theta_2$ 2 vertically: $T_1 \cos \theta_1 + T_2 \cos \theta_2 = W$.

The parallelogram of forces

In Figure 1, the resultant of the three forces is zero. Therefore, the resultant of any *two* of the forces is equal and opposite to the *third* force. We can show this by using the parallelogram of forces rule to draw a scale diagram as follows:

1 Measure the angles θ_1 and θ_2 between each of the upper strings and the lower string which is vertical.

2 Draw a scale diagram of a parallelogram as shown in Figure 2, using the two force vectors T_1 and T_2 as adjacent sides.

3 The resultant of T_1 and T_2 is equal and opposite to W. The resultant is the diagonal of the parallelogram *between* the two force vectors.

Summary questions

1 A point object of weight 3.2 N is acted on by a horizontal force of 5.8 N.
   a Calculate the resultant of these two forces. *(3 marks)*
   b Determine the magnitude and direction of a third force acting on the object for it to be in equilibrium. *(1 mark)*

2 A small object of weight 8.6 N is at rest on a rough slope, which is at an angle of 40° to the horizontal, as shown in Figure 3. Calculate:
   a the frictional force F on the object
   b the support force S from the slope on the object. *(3 marks)*

3 A string fixed at each end at different heights a fixed distance apart supports a stationary object of weight 2.0 N. Calculate the tension in each section of the string when the object is stationary and the angles of each section of the string to the horizontal are 35° and 30°. *(4 marks)*

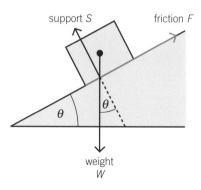

▲ **Figure 3**

6.3 The principle of moments

Specification reference: 3.4.1.2

The **moment** of a force about any point is defined as the force × the perpendicular distance from the line of action of the force to the point.

The unit of the moment of a force is the newton metre (N m).

For a force F acting along a line of action at perpendicular distance d from a certain point, **the moment of the force = $F \times d$.**

The principle of moments

If a body acted on by forces is in equilibrium,

the sum of the clockwise moments about any fixed point $=$ the sum of the anticlockwise moments about that point

Consider a uniform metre rule balanced at its centre, as in Figure 2.

- Weight W_1 provides an anticlockwise moment about the pivot = W_1d_1.
- Weight W_2 provides a clockwise moment about the pivot = W_2d_2.

For equilibrium, applying the principle of moments,

$$W_1d_1 = W_2d_2$$

Note: The support force $S = W_1 + W_2$

Centre of mass

The **centre of mass** of a body is the point through which a single force on the body has no turning effect.

Figure 3 shows a metre rule that has been balanced off-centre on a knife-edge by adjusting the position of a known weight W_1.

- The known weight W_1 provides an anticlockwise moment about the pivot = W_1d_1.
- The weight of the rule W_0 provides a clockwise moment = W_0d_0.

Applying the principle of moments therefore gives $W_0d_0 = W_1d_1$.

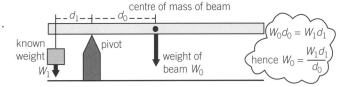

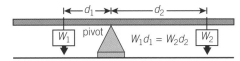

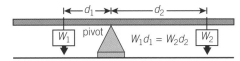

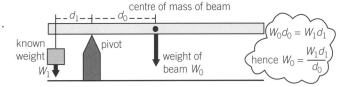

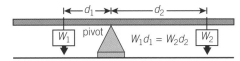

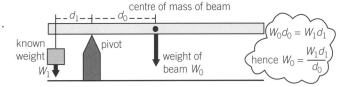

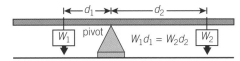

▲ **Figure 1** *A turning force*

▲ **Figure 2** *Using the principle of moments*

▲ **Figure 3** *Finding the weight of a beam*

Distance d is the **perpendicular** distance from the line of action of the force to the point.

Summary questions

1. A metre rule, pivoted at its centre of mass, supports a 2.0 N weight at its 10.0 cm mark, a 3.0 N weight at its 25 cm mark, and a weight W at its 90 cm mark.
   a. Sketch a diagram to represent this situation. *(1 mark)*
   b. Calculate the weight W. *(3 marks)*

2. A child of weight 200 N sits at one end of a seesaw at a distance of 1.2 m from the pivot at the centre. The seesaw is balanced by a second child of weight 120 N sitting on it at a distance of 0.8 m from the centre and an adult holding the seesaw at one end 0.4 m from the second child. Calculate the force exerted by the adult on the seesaw. *(3 marks)*

3. A uniform metre rule of weight 1.2 N is balanced horizontally on a horizontal knife-edge at its 250 mm mark. The ruler supports a 2.5 N weight at its 150 mm mark and is supported by a vertical string attached to a stand at one end and to the ruler at the 950 mm mark at the other end. Sketch the arrangement and calculate the tension in the string. *(4 marks)*

Revision tip

When there are several unknown forces, take moments about a point through which one of the unknown forces acts. This will simplify your calculation.

6.4 More on moments
6.5 Stability

Specification reference: 3.4.1.1; 3.4.1.2

Support forces

When an object in equilibrium is supported at one point only, the support force on the object is equal and opposite to the total downward force acting on the object.

Consider a uniform beam supported on two pillars X and Y, which are at distance D apart. The weight of the beam is shared between the two pillars according to how far the beam's centre of mass is from each pillar. For example:

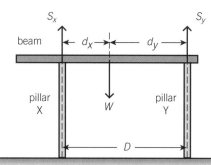

▲ **Figure 1** *A two-support problem*

- Suppose the centre of mass of the beam is at distance d_x from pillar X and distance d_y from pillar Y, as shown in Figure 1, then taking moments about where X is in contact with the beam gives,

  $S_y D = W d_x$, where S_y is the support force from pillar Y.

  Therefore, $S_y = \dfrac{W d_x}{D_1}$.

Note: To determine S_x, use $S_x + S_y = W$.

Couples

A **couple** is a pair of equal and opposite forces acting on a body, but not along the same line. Figure 2 shows a couple acting on a beam. The couple turns or tries to turn the beam.

The moment of a couple = force × perpendicular distance between the lines of action of the forces.

The moment of a couple about any point is always the same. In Figure 2 the total moment about an arbitrary point P along the beam $= Fx + F(d - x)$ $= Fx + Fd - Fx = Fd$.

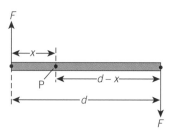

▲ **Figure 2** *A couple*

Stable and unstable equilibrium

- If a body in **stable equilibrium** is displaced then released, it returns to its equilibrium position. This happens because when it is displaced, its centre of mass is not directly below the point of support. So the weight of the object returns it to equilibrium.

- If an object in **unstable equilibrium** is displaced slightly from equilibrium then released, it will move away from equilibrium. The reason is that when the object is displaced slightly, the centre of mass is no longer above the point of support. The object's weight therefore acts to turn it further from the equilibrium position.

Tilting and toppling

Tilting is where an object at rest on a surface is acted on by a force that makes it turn about a point or a line where it is in contact with the surface. A tilted object will topple over if it is tilted too far.

Summary questions

1 A uniform beam of length 2.0 m and weight 120 N rests horizontally on two supports X and Y, 0.20 m and 0.25 m respectively from each end. Sketch the arrangement and calculate the support force on the beam at each end.   (5 marks)
   Hint: Mark the relevant distances as well as the forces on your diagram.

2 A bridge crane consists of a horizontal span of weight 6400 N and length 11.0 m fixed at each end to vertical pillars P and Q as shown in Figure 3. The span supports a load of 500 N at a distance of 4.0 m from pillar P.
   Calculate the support force on the span from each pillar.   (4 marks)

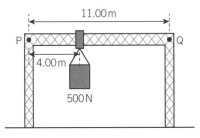

▲ **Figure 3**

3 A horizontal uniform diving board AB of weight 200 N and of length 2.50 m overhangs the edge of the swimming pool by 1.50 m. The board is bolted to the ground at the edge of the pool P and at end A on the ground. A diver of weight 640 N stands on the board over the pool at end B.
   a Sketch a diagram of the arrangement and show the direction of the forces on the board.   (1 mark)
   b i Show that the force of the bolt at A on the board is 1010 N.
                                                               (3 marks)
      ii Calculate the force of the bolt at P on the board.   (1 mark)

4 Explain why side winds on a high-sided vehicle can cause it to topple over.   (2 marks)

5 A filing cabinet is usually designed so only one drawer can be pulled out at any time. Explain why the cabinet would be unstable if more than one drawer was pulled out at the same time.   (3 marks)

6 Figure 4 shows a vehicle being driven slowly across a slope. The distance between the wheels on each side of the vehicle is 1.7 m and its centre of mass is 0.9 m from the ground. Calculate the maximum angle of a slope on which the vehicle would not topple over.   (3 marks)

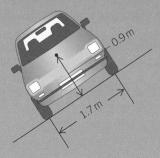

▲ **Figure 4**

6.6 Equilibrium rules
6.7 Statics calculations
Specification reference: 3.4.1.1; 3.4.1.2

Free body force diagrams

When two objects interact, they always exert equal and opposite forces on one another. A **free body force diagram** shows only the forces acting on one object.

The triangle of forces

For an object acted on by three forces to be in equilibrium, the three forces must give an overall resultant of zero. The three forces as vectors should form a triangle. In other words, for three forces F_1, F_2, and F_3 to give zero resultant:

their vector sum $F_1 + F_2 + F_3 = 0$

- Their lines of action must intersect at the *same* point, otherwise the object cannot be in equilibrium, as the forces will have a net turning effect.

- A scale diagram of the triangle of forces can be drawn to find an unknown force or angle, given the other forces and angles in the triangle.

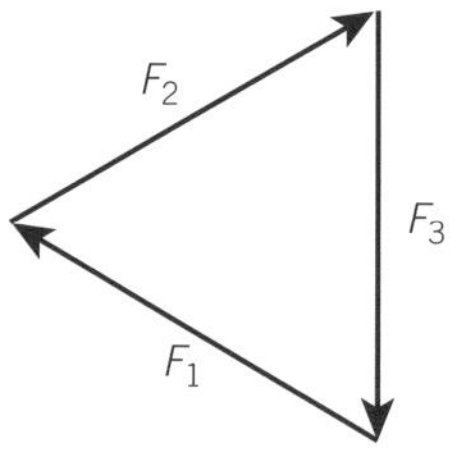

The conditions for equilibrium of a body

For a body in equilibrium:

1 The resultant force must be zero. If there are only three forces, they must form a closed triangle.

2 The principle of moments must apply (i.e., the moments of the forces about the same point must balance out).

▲ **Figure 1** *The triangle of forces for a point object*

Synoptic link

See Topics 6.3, The principle of moments, and 6.4, More on moments, again if necessary.

Summary questions

1 A uniform plank of length 4.0 m and of weight 180 N rests horizontally on two bricks X and Y at 0.10 m from either end. A child C of weight 150 N stands at a distance of 0.90 m from one end.
   a Draw a free body force diagram of the forces acting on the plank. *(1 mark)*
   b Calculate the support forces acting on the plank from the supporting bricks. *(4 marks)*

2 A uniform ladder of weight 220 N rests against a smooth vertical wall at an angle of 20° and with its lower end on flat ground. The force on the ladder due to the wall is perpendicular to the wall.
   a Draw a free body force diagram of the forces acting on the ladder. *(1 mark)*
   b Determine the magnitude and direction of the force of the ground on the ladder. *(4 marks)*

3 A rectangular picture of weight 21 N is supported on a wall by a cord attached to the top corners of the picture which hangs over a wall hook directly above the midpoint of the picture. Each section of the cord is at an angle of 25° to the top edge of the picture, which is horizontal. The centre of mass of the picture is at its geometrical centre.
   a Draw a free body force diagram of the forces acting on the picture. *(1 mark)*
   b Calculate the tension in each section of the cord. *(3 marks)*

Chapter 6 Practice questions

1 A point object in equilbrium is acted on by a 3 N force, a 5 N force, and a 6 N force, which is horizontal.

    a Draw a triangle of forces diagram to determine the angle of each of the lines of action of the other forces to the 6 N force.   (*3 marks*)

    b What is the resultant force on the object if the 5 N force is removed?

        (*1 mark*)

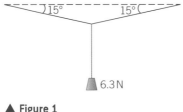

▲ Figure 1

2 An object of weight 6.3 N hangs on the end of a cord, which is attached to the midpoint of a wire stretched between two points on the same horizontal level, as shown in Figure 1. Each half of the wire is at 15° to the horizontal. Calculate the tension in each half of the wire.   (*4 marks*)

3 A ship is towed at constant velocity by two tugboats which pull the ship with forces of 8.0 kN and 7.2 kN respectively. The angle between the tugboat cables is 45°, as shown in Figure 2.

    a Determine the angle between each cable and the direction of motion of the ship.

       (Hint: draw a parallelogram of forces and use it to establish the direction of motion on the diagram.)   (*3 marks*)

    b Calculate:

      i   the magnitude of the resultant of the two tugboat forces

      ii   the drag force on the ship.   (*3 marks*)

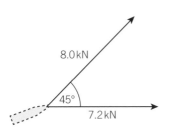

▲ Figure 2

4 A uniform metre rule of weight 1.20 N is pivoted on a metal rod at its 350 mm mark and is also supported in a horizontal position by a string inclined at 60° to the horizontal, as shown in Figure 3. One end of the string is tied to a stand and the other end is attached to the 950 mm mark on the ruler.

    a Sketch the arrangement (without showing the stand) and calculate the tension in the string.   (*3 marks*)

    b Calculate the support force on the ruler from the pivot as in Figure 3.   (*4 marks*)

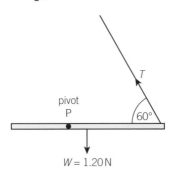

▲ Figure 3

5 A uniform plank of weight 350 N and of length 4.00 m is supported horizontally on two horizontal scaffolding tubes X and Y at 0.20 m from each end. A person P of weight 700 N stands on the plank at a distance of 1.50 m from end X.

    a Sketch a free body force diagram of the beam.   (*1 mark*)

    b Calculate the support force on the beam from each scaffolding tube.

        (*4 marks*)

6 A uniform curtain pole of weight 22 N and of length 2.80 m is supported horizontally by two wall-mounted brackets A and B, which are 0.05 m from each end. The pole supports a pair of curtains of total weight 86 N.

    a Calculate the support force of each bracket on the curtain pole when the curtains are drawn along the full length of the pole between the brackets,   (*3 marks*)

    b With the curtains initially as in **a**, the curtain nearest A is pulled back to A without moving the other curtain. Describe how the support force at A changes as the curtain nearest A is pulled back to A without moving the other curtain.   (*2 marks*)

    c Calculate the support force on bracket A when the curtain nearest A has been pulled back to A without moving the other curtain.   (*5 marks*)

       (Hint: Draw a free body diagram to show the forces on the curtain pole assuming the centre of mass of each curtain is at the centre of the curtain.)

7 A wire cable of length 6.5 m is fixed at its ends to two clamps X and Y which are about 6.3 m apart at the same height. The cable supports an object of weight 30 N at a point P as shown in Figure 4. At this position, section XP is inclined at an angle of 20° below XY and the other section YP is inclined an angle of 15° below XY.

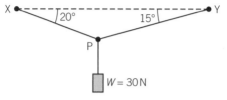

▲ Figure 4

a Show that the tension in section XP is 1.03 times the tension in section YP.   (2 marks)

b Calculate the tension in each section of the cable.   (4 marks)

8 A suitcase has a small pair of wheels at one of its lower corners and an extended handle on its top side. Figure 5 shows the suitcase when it is upright and stationary on a flat surface.

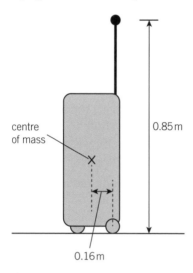

▲ Figure 5

The suitcase has a weight of 220 N and its centre of mass is a horizontal distance of 0.16 m from the axis of the wheels. The handle of the suitcase is 0.85 m directly above the axis of the wheels.

a Calculate the horizontal force that must be applied to the handle to raise one side of the suitcase off the ground.   (2 marks)

b Explain why the position of the suitcase is unstable when it is tilted so its centre of mass is directly above the wheel axis.   (2 marks)

c If the suitcase is tilted more than in part **b**, explain why an upward force on the handle is necessary to keep it stationary.   (2 marks)

7.1 Speed and velocity

Specification reference: 3.4.1.3

Speed

Speed and distance are scalar quantities. Velocity and displacement are vector quantities.

The unit of speed and of velocity is the metre per second ($m\,s^{-1}$).

Motion at constant speed

For an object which travels distance s in time t at constant speed:

- speed $v = \dfrac{s}{t}$
- a graph of distance against time is a straight line with a constant gradient which is equal to the speed of the object.

For an object moving round a circle of radius r at constant speed, its speed $v = \dfrac{2\pi r}{T}$ where T is the time to move round once and $2\pi r$ is the circumference of the circle.

Motion at changing speed

For an object moving at changing speed:

- over a distance s in time t, its average speed $= \dfrac{s}{t}$
- its speed v at any instant is given by $v = \dfrac{\Delta s}{\Delta t}$ where Δs is the distance it travels in a short time interval Δt
- a graph of distance against time has a gradient that changes with time. The gradient of the line at any point can be found by drawing a tangent to the line at that point and then measuring the gradient of the tangent. See Figure 1.

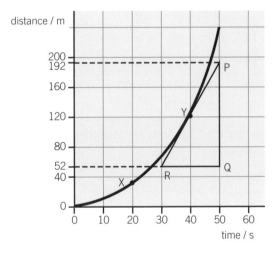

speed at $Y = \dfrac{PQ}{QR} = \dfrac{192 - 52}{20}$

$= 7\,m\,s^{-1}$

▲ **Figure 1**

Velocity

An object moving at **constant velocity** moves at the same speed without changing its direction of motion. If an object changes its direction of motion or its speed or both, its velocity changes.

Figure 2 shows a displacement–time graph for an object thrown into the air. Upward directions are considered as positive and downward directions as negative. Initially, the object has zero displacement and a positive velocity.

During its ascent, its displacement and velocity (= gradient of the displacement–time graph) are both positive.

At maximum height, its displacement is at its maximum value and its velocity is zero.

Key terms

Displacement is defined as the distance in a given direction.

Speed is defined as the change of distance per unit time.

Velocity is defined as change of displacement per unit time. In other words, velocity is speed in a given direction.

Revision tip

$1\,km = 1000\,m$, $1\,hour = 3600\,s$, $108\,km\,h^{-1} = 30\,m\,s^{-1}$

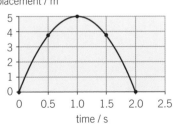

▲ **Figure 2** *A displacement–time graph*

During its descent, its velocity is negative and its displacement is positive until it returns to its initial position.

Displacement–time graphs are covered in more detail in Topic 7.5, Motion graphs.

Summary questions

1 A vehicle joins a motorway and travels a distance of 42 km in 25 minutes and a further distance of 20 km at a speed of $20 \, m \, s^{-1}$ before leaving the motorway.
  a Calculate the speed of the vehicle in the first part of the journey. *(1 mark)*
  b Calculate the average speed of the vehicle for the whole journey. *(3 marks)*

2 The Earth has an equatorial radius of 6 360 km and it rotates about its axis once every 24 hours.
  a Calculate the speed of an object on the Earth's surface at the equator. *(2 marks)*
  b A satellite orbits the Earth once every two hours in a circular orbit that passes over the Earth's poles. The satellite passes directly over a certain point P on the equator.
   i State the time taken to the next time the satellite passes directly over the equator in the same direction. *(1 mark)*

   ii Calculate the distance along the equator from P to the point Q on the equator directly under the satellite at the time calculated in i. *(3 marks)*

3 a Explain the difference between speed and velocity. *(2 marks)*
  b The table gives the distance moved from rest by an aircraft on a runway at different times during take-off.

Displacement / m	0	200	400	800	1200	1600	2000
Time / s	0	10	20	28	35	40	46

   i Use the data in the table to plot a displacement–time graph. *(3 marks)*

   ii Determine the speed of the aircraft at 30 s. *(2 marks)*

7.2 Acceleration

Specification reference: 3.4.1.3

Uniform acceleration

Uniform acceleration is where the velocity of an object moving along a straight line changes at a constant rate. In other words, the acceleration is constant. Consider an object that accelerates uniformly from velocity u to velocity v in time t along a straight line, as shown in Figure 1.

The acceleration, a, of the object $= \dfrac{\text{change of velocity}}{\text{time taken}} = \dfrac{v - u}{t}$

Rearranging this equation gives $v = u + at$

Non-uniform acceleration

Non-uniform acceleration is where the direction of motion of an object changes, or its speed changes, at a varying rate. Figure 2 shows how the velocity of an object increases for an object moving along a straight line with an increasing acceleration. This can be seen directly from the graph because the gradient increases with time (the graph becomes steeper and steeper) and the gradient represents the acceleration. The acceleration at any point is the gradient of the tangent to the curve at that point.

acceleration = gradient of the line on the velocity–time graph.

Summary questions

1 A car travelling at a velocity of 26 m s^{-1} brakes sharply to a standstill in 6.7 s. Calculate its deceleration, assuming its velocity decreases uniformly. *(2 marks)*

2 An electron in a vacuum tube is accelerated in a straight line from rest to a velocity of 1.8×10^7 m s^{-1} in a time of 51 ns. Calculate the acceleration of the electron. *(2 marks)*

3 Figure 3 shows the velocity of an object released at the surface of a liquid in a vertical tube.
   a Describe how: **i** the velocity of the object changed with time,
      ii the acceleration of the object changed with time. *(2 marks)*
   b Use the graph in Figure 3 to determine the acceleration of the object 0.50 s after it was released. *(2 marks)*

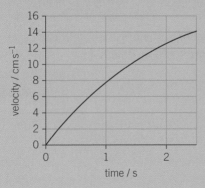

▲ Figure 3

Revision tip

- The unit of acceleration is the metre per second per second (m s^{-2}).
- Acceleration is a vector quantity.
- For a moving object that does not change direction, its acceleration at any point can be worked out from its rate of change of velocity, because there is no change of direction.

Key term

Acceleration is defined as change of velocity per unit time.

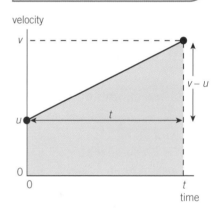

▲ Figure 1

Revision tip

Use of the equation $a = \dfrac{v - u}{t}$ for *non-uniform* acceleration gives an average value for the acceleration during time t.

Revision tip

Remember that acceleration is a vector quantity. It is dependent on changes in both speed and direction.

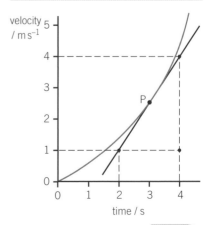

▲ Figure 2

7.3 Motion along a straight line at constant acceleration

Specification reference: 3.4.1.3

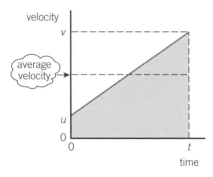

▲ Figure 1

Revision tip

Remember the acceleration must be constant for these equations to apply.

Worked example

A driver of a vehicle travelling at a speed of 31.0 m s⁻¹ on a motorway brakes to a speed of 8.0 m s⁻¹ in a distance of 105 m. Calculate the deceleration of the vehicle.

Solution

$u = 31.0\,\mathrm{m\,s^{-1}}, v = 8.0\,\mathrm{m\,s^{-1}},$
$s = 105\,\mathrm{m}, a = ?$

To find a, use $v^2 = u^2 + 2as$

Rearranging this equation gives
$2as = v^2 - u^2$

Therefore
$a = \dfrac{v^2 - u^2}{2s} = \dfrac{8.0^2 - 31.0^2}{2 \times 105} = -4.3\,\mathrm{m\,s^{-2}}$
The deceleration is 4.3 m s⁻².

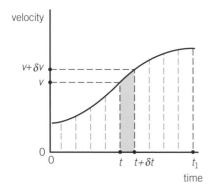

▲ Figure 2

The dynamics equations for constant acceleration

Consider an object that accelerates uniformly without change of direction from initial velocity u to final velocity v and displacement s in time t. Figure 1 shows how its velocity changes with time. The following equations may be used to solve problems involving uniform (i.e., constant) acceleration.

1 $v = u + at$
2 $s = \dfrac{(u + v)t}{2}$
3 $s = ut + \dfrac{1}{2}at^2$
4 $v^2 = u^2 + 2as$

These four equations are used in any situation where the acceleration is constant. Always begin a *suvat* calculation by identifying which three variables are known, then choose the equation which has these three variables in it together with the variable to be calculated.

Using a velocity–time graph to find the displacement

1 For an object moving at *constant* acceleration, a, from initial velocity u to velocity v at time t, as shown in Figure 1, its displacement, $s = \dfrac{(u + v)t}{2}$. This is represented on the graph by the area under the line between the start and time t. This is a trapezium which has an area (= average height × base) corresponding to $\dfrac{1}{2}(u + v)t$, which is equal to the displacement.

2 Consider an object moving with a *changing* acceleration, as shown in Figure 2. For a velocity change from v to $v + \delta v$ in a short time interval, its displacement $\delta s = v\,\delta t$. This is represented on the graph by the area of the shaded strip under the line. By considering the whole area under the line in strips of similar width, the total displacement from the start to time t is represented by the sum of the area of every strip, which is therefore the total area under the line.

Whatever the shape of the line of a velocity–time graph,

displacement = area under the line of a velocity–time graph.

Summary questions

1 A vehicle accelerates uniformly from rest along a straight road and reaches a velocity of 28.0 m s⁻¹ after 170 m. Calculate:
   a its acceleration   (2 marks)
   b the time taken.   (2 marks)

2 An aircraft lands on a runway at a velocity of 86 m s⁻¹ and brakes to a halt 28 s later. Calculate:
   a the distance travelled in this time   (2 marks)
   b the deceleration of the aircraft.   (2 marks)

3 Figure 3 in Topic 7.2, Acceleration, shows how the velocity of an object falling in a liquid changes with time after being released at the liquid surface. Estimate the distance fallen by the object in 2.5 s.   (2 marks)

7.4 Free fall
Specification reference: 3.4.1.3

For a falling object on which no external forces act, apart from the force of gravity, its acceleration is constant and is referred to as the **acceleration of free fall**, g. Accurate measurements give a value of $9.81\,\text{m}\,\text{s}^{-2}$ near the Earth's surface.

The 'suvat' equations from Topic 8.3, Terminal speed may be applied to any free fall situation where air resistance is negligible using the value of g as the acceleration. As a general rule, apply the direction code $+$ for **upwards** and $-$ for **downwards** when values are inserted into the *suvat* equations.

Straight line graphs

The value of g can be determined by measuring the distance which a suitable object falls through in different times. In time t, the distance fallen $s = ut + \frac{1}{2}at^2$, where u = the initial speed and a = acceleration.

If the object is released from rest and s is measured from the point of release, then $u = 0$ so $s = \frac{1}{2}at^2$. A graph of s against t^2 should give a straight line through the origin which has a gradient $\frac{1}{2}a$. Hence, $g = 2 \times$ the gradient of the graph.

If the distances are measured from a point below where the object is released, then $u \neq 0$ so a graph of $s = ut + \frac{1}{2}at^2$ would not be a straight line. Dividing the equation $s = ut + \frac{1}{2}at^2$ by t gives $\frac{s}{t} = u + \frac{1}{2}at$. Therefore, a graph of $\frac{s}{t}$ against t can be plotted to give a straight line with a y-intercept equal to u and a gradient $\frac{1}{2}a$.

Worked example
$g = 9.81\,\text{m}\,\text{s}^{-2}$

A tennis ball thrown vertically upwards reaches a height of 14.0 m then descends to the ground below. Calculate: **a** the initial speed of the ball, **b** its displacement 3.00 s after it was thrown into the air.

Solution

a $u = ?$, $v = 0$ at maximum height, $s = 14.0\,\text{m}$, $a = -9.81\,\text{m}\,\text{s}^{-2}$, $t = ?$

   To find u, use $v^2 = u^2 + 2as$. Note that a is negative because the direction of g is downwards.

   Rearranging this equation gives $u^2 = v^2 - 2as = 0 - (2 \times -9.81 \times 14.0) = 275\,\text{m}^2\,\text{s}^{-2}$

   Hence, $u = \pm 16.6\,\text{m}\,\text{s}^{-1}$ and since its initial velocity was upwards, $u = 16.6\,\text{m}\,\text{s}^{-1}$

b To find s at 3.00 s, use $s = ut + \frac{1}{2}at^2 = (16.6 \times 3.00) + (0.5 \times -9.81 \times 3.00^2) = 5.66\,\text{m}$

   The position of the ball 3.00 s after being released was 5.66 m above the point at which it left the thrower's hand.

Revision tip
The general equation for a straight line graph is $y = mx + c$ where m is the gradient of the line and c is the y-intercept.

7.5 Motion graphs
7.6 More calculations on motion along a straight line

Specification reference: 3.4.1.3

Revision tip

Displacement is distance in a given direction from a certain point. Velocity is speed in a given direction.

Distance–time and displacement–time graphs

The gradient of the line:

- on a distance–time graph represents the speed of the object
- on a displacement–time graph represents the velocity of the object
  - a positive gradient means the velocity at that point is in the direction assigned as the positive direction
  - a negative gradient means the velocity is in the opposite direction.

The difference between a speed–time graph and a velocity–time graph

The gradient of the line:

- on a speed–time graph represents the magnitude of the acceleration
- on a velocity–time graph represents the magnitude and direction of the object's acceleration.

The area under the line:

- of a speed–time graph represents the distance travelled
- of a velocity–time graph represents the displacement of the object from its starting position. The area between the positive section of the line and the time axis represents a positive displacement. The area under the negative section of the line and the time axis represents a negative displacement.

Worked example

Figure 1 shows the velocity–time graph for a ball that is released from rest at a height h_1 above a concrete floor, hits the floor at time t_1, and rebounds to a maximum height h_2 at time t_2.

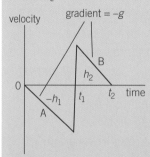

▲ Figure 1

a Explain why the graph line has two sections A and B with the same gradient.

b Explain why the line is almost vertical between A and B.

c Explain why the area between the time axis and each section is different.

Solution

a In both sections, the acceleration of the ball is due to gravity only and therefore the gradient of both sections is the same.

b The ball undergoes a large upward force in a very short time when it hits the floor at time t_1, causing a sudden reversal of its velocity.

c The displacement in each section = the area between the line and the time axis. The displacement is $-h_1$ for A and $+h_2$ for B. Since height h_2 is less than height h_1, the area between B and the time axis is less than the area between A and the time axis.

Using the equations for constant acceleration

The equations for motion at constant acceleration, a, are only valid when the acceleration of an object does not change.

Note:

• If a certain direction is assigned as the positive direction, then any calculated value that is negative is in the opposite direction.

• If the motion of an object is in stages where each stage has a different constant acceleration, the equations from Topic 7.3, Motion along a straight line at constant acceleration, may be used at each stage. The final velocity at the end of each stage is the initial velocity at the start of the next stage.

Synoptic link

Notice in Figure 1 that h_2 is less than h_1 because the ball loses some energy due to the impact when it rebounds. So its potential energy relative to the floor at maximum height after the rebound is less than its initial potential energy. See Topic 10.2, Kinetic energy and potential energy, for more about potential energy.

Summary questions

$g = 9.81\ \mathrm{m\,s^{-2}}$

1 A bungee jumper jumps from a platform at O and is in free fall until she reaches a point A when the rope attached to her starts to slow her descent and reduces her speed momentarily to zero at B. She then ascends and completes the jump.
  a Sketch a velocity–time graph for her descent from O to B. Label the points A and B on the graph. *(2 marks)*
  b Describe how her acceleration changed from O to B. *(3 marks)*

2 A small object released from rest above a level surface falls freely and hits the surface at a speed of $2.4\ \mathrm{m\,s^{-1}}$.
  a i Calculate its time of descent.
    ii Sketch a graph to show how its velocity changes with time during its descent. *(4 marks)*
  b Calculate the distance fallen by the object. *(2 marks)*

3 A rocket launched vertically from rest on the ground accelerated at a constant acceleration of $3.70\ \mathrm{m\,s^{-2}}$ for 30 seconds when the rocket engines were switched off. The rocket continued to ascend and then fell to the ground.
  a Calculate the velocity of the rocket and its displacement after 30 seconds. *(4 marks)*
  b i Calculate how long the rocket took to reach its maximum height after the rockets were switched off. *(2 marks)*
    ii Sketch a graph to show how the velocity of the rocket changed with time during its ascent to maximum height. *(3 marks)*
  c Calculate the velocity of the rocket just before it hit the ground. *(3 marks)*

A **projectile** is acted upon only by the force of gravity. Three key principles apply to all projectiles:

- The acceleration of the object is always equal to g and is always downwards because the force of gravity acts downwards. The acceleration therefore only affects the vertical motion of the object.

- The horizontal velocity of the object is constant because the acceleration of the object does not have a horizontal component.

- The motions in the horizontal and vertical directions are independent of each other.

1 Vertical projection

An object that moves vertically has no horizontal motion. See Topic 7.4, Free fall, for notes on vertical projection.

2 Horizontal projection

An object projected horizontally falls in a parabolic arc towards the ground as shown in Figure 1. If its initial velocity is U, then at time t after projection:

- Its displacement has:

  - a *horizontal* component $x = Ut$ (because it moves horizontally at a constant speed)

  - a *vertical* component $y = \frac{1}{2}gt^2$ (because its acceleration g is vertically downwards).

- Its velocity has a horizontal component $v_x = U$, and a vertical component $v_y = -gt$.

- At time t, its speed $= (v_x^2 + v_y^2)^{\frac{1}{2}}$ and its velocity direction is at angle θ to the horizontal where $\tan\theta = \dfrac{v_y}{v_x}$.

object projected horizontally at speed U

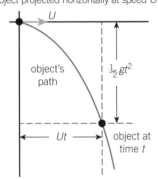

▲ **Figure 1** *Horizontal projection*

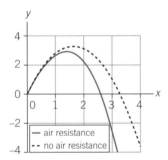

▲ **Figure 2** *Projectile paths*

3 Projection at angle θ above the horizontal

An object that is projected in a direction above the horizontal rises and falls in a parabolic arc. Because its acceleration, g, is vertically downwards, the horizontal component of its velocity is constant and the vertical component changes at a constant rate equal to $-g$. The dotted arc in Figure 2 shows the path of such a projectile when air resistance is negligible.

Worked example

$g = 9.81\,\mathrm{m\,s^{-2}}$

An object is projected horizontally from the top of a tower of height 25.0 m and it hits the flat ground below 40.0 m away from the base of the tower. Calculate the speed of projection of the object.

Solution

Consider the vertical motion: $y = -25\,\mathrm{m}$, $u_y = 0$, $a_y = -9.81\,\mathrm{m\,s^{-2}}$ ($-$ for downwards)

Rearranging gives $y = \frac{1}{2}a_y t^2$ gives $t^2 = \dfrac{2y}{a_y} = \dfrac{2 \times -25.0}{-9.81} = 5.10\,\mathrm{s^2}$ so $t = 2.26\,\mathrm{s}$

Consider the horizontal motion: $x = 40.0\,\mathrm{m}$, $a_x = 0$, $t = 2.26\,\mathrm{s}$

Rearranging $x = Ut$ gives $U = \dfrac{x}{t} = \dfrac{40.0}{2.26} = 17.7\,\mathrm{m\,s^{-1}}$

The effects of air resistance

A projectile moving through air experiences a **drag force** that opposes the motion of the projectile and which increases as the projectile's speed increases. The lower arc in Figure 2 shows that drag on a projectile:

- reduces the horizontal speed of the projectile and its range
- makes its descent steeper than its ascent
- reduces the maximum height if its initial direction is above the horizontal.

The shape of the projectile is important because it affects the drag force and may also cause a **lift force** in the same way as the cross-sectional shape of an aircraft wing creates a lift force. A spinning ball moving through the air also experiences a force due to the same effect. However, this force can be downwards, upwards, or sideways, depending on how the ball is made to spin.

Projectile-like motion

Any form of motion where an object experiences a constant acceleration in a different direction to its velocity will be like projectile motion. For example:

- The path of a ball rolling across an inclined board will be a projectile path. Its path curves down the board because the object is subjected to constant acceleration acting down the board, and its initial velocity is across the board.
- The path of a beam of electrons directed between two oppositely charged parallel plates is similar in shape. This is because each electron in the beam experiences a constant acceleration towards the positive plate and its initial velocity is parallel to the plates.

Summary questions

$g = 9.81 \, \text{m s}^{-2}$

1 An object is projected horizontally at a speed of $16 \, \text{m s}^{-1}$ into the sea from a cliff top and falls into the sea 3.1 s later. Calculate:
   a the height of the cliff top above the sea (*2 marks*)
   b how far it travels horizontally (*1 mark*)
   c its impact speed. (*3 marks*)

2 A parcel was released from an aircraft travelling horizontally at a constant velocity of $110 \, \text{m s}^{-1}$ above level ground. The parcel hit the ground 7.5 s later.
   a Calculate the height of the aircraft above the ground. Assume the drag force on the parcel is negligible. (*2 marks*)
   b Discuss whether or not the aircraft was directly above the parcel when the parcel hit the ground. (*3 marks*)

3 An archer on a flat field fired an arrow at a speed of $29.0 \, \text{m s}^{-1}$ at an angle of $15.0°$ above the horizontal.
   a Calculate the horizontal and vertical components of velocity of the arrow at the instant it was released. (*1 mark*)
   b The arrow was 2.60 m above the ground when it was fired. By considering its vertical motion, calculate:
      i the vertical component of the arrow's velocity when it hit the ground (*2 marks*)
      ii how long the arrow took to fall to the ground. (*2 marks*)
   c Calculate the horizontal displacement from the archer to the point where the arrow hit the ground. (*1 mark*)

$g = 9.81\,\mathrm{m\,s^{-2}}$

1   A car decelerated uniformly to rest from a speed of $98\,\mathrm{km\,h^{-1}}$ in a distance of 73 m.

    a Calculate the deceleration of the car in $\mathrm{m\,s^{-2}}$. *(2 marks)*

    b Calculate the time taken by the car whilst decelerating. *(2 marks)*

2   Figure 1 shows how the velocity of an aircraft changes when it lands on a runway.

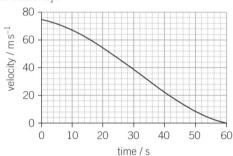

▲ **Figure 1**

    a Determine the maximum deceleration of the aircraft. *(3 marks)*

    b Estimate the distance it travelled along the runway until it stopped. *(2 marks)*

3   In a 100 m race, athlete X accelerated uniformly for 1.20 s from rest to a speed of $9.60\,\mathrm{m\,s^{-1}}$ and completed the race at that speed.

    a **i** Calculate the distance travelled by the athlete in the first 1.20 s.

        ii Calculate the total time taken by the athlete to run the race. *(4 marks)*

    b A second athlete Y starting at the same time, accelerated uniformly for 1.30 s to a certain speed and maintained that speed to finish the race at the same time as X. Determine the maximum speed of Y. *(3 marks)*

4   A rail wagon moving at a speed of $1.80\,\mathrm{m\,s^{-1}}$ on a level track reached a steady incline which slowed it down to rest in 12.0 s and caused it to reverse.

    a Calculate its rate of change of velocity on the incline. *(2 marks)*

    b Calculate:

        i its velocity after 16.0 s

        ii its position on the incline 16.0 s after it reached the incline. *(4 marks)*

5   A cyclist accelerated uniformly from rest at P to a speed of $8.0\,\mathrm{m\,s^{-1}}$ in 20.0 s at Q then braked at uniform deceleration to a halt in a distance of 20.0 m at R. She then turned her cycle round in 10.0 seconds and then cycled at a speed of $4.0\,\mathrm{m\,s^{-1}}$ for 5.0 s in the opposite direction to the original direction before she stopped at S.

Figure 2 shows the route which was on level ground

▲ **Figure 2**

    a **i** Calculate the distance she moved from P to Q.

        ii Calculate how long she took to cycle from Q to R. *(2 marks)*

    b Plot a velocity–time graph of her journey from P to S. Label the points on your graph corresponding to P, Q, R, and S. *(2 marks)*

c i Calculate the acceleration of the cyclist when she travelled from P to Q.

    ii Calculate her displacement from P to S. *(3 marks)*

6 A stone hit the water in a river 2.0 s after being thrown horizontally at a speed of 14 m s^{-1} from a bridge above the river.

    a Calculate the distance from the bridge to the point where the stone hit the water. *(4 marks)*

    b State and explain how the horizontal displacement would have differed if the stone had been thrown horizontally with a lesser speed. *(5 marks)*

7 An object was released from a hot air balloon when it was travelling at a speed of 2.60 m s^{-1} in a direction 30.0° above the horizontal as shown in Figure 3.

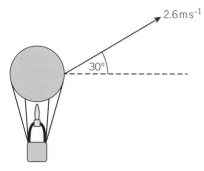

▲ **Figure 3**

    a Calculate the horizontal and vertical components of velocity of the object at the instant it was released. *(2 marks)*

    b The object took 3.90 s to fall to the ground. Calculate:

        i the vertical distance fallen by the object from the point of release

        ii the horizontal distance travelled by the object from the point of release to where it hit the ground. *(3 marks)*

8 A javelin thrown on a level field by an athlete from a height of 1.80 m above the ground hit the ground 85.5 m away 4.24 s after it was released.

    a i Calculate the horizontal component of the javelin's velocity.

        ii Calculate the vertical component of the javelin's initial velocity. *(4 marks)*

    b i Show that the speed at which the javelin was thrown was about 29 m s^{-1}.

        ii Calculate the angle θ above the horizontal at which the javelin was thrown. *(2 marks)*

8.1 Force and acceleration

Specification reference: 3.4.1.5

Worked example

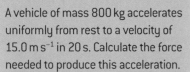

A vehicle of mass 800 kg accelerates uniformly from rest to a velocity of 15.0 m s^{-1} in 20 s. Calculate the force needed to produce this acceleration.

Solution

$$\text{acceleration } a = \frac{v - u}{t} = \frac{15 - 0}{20}$$

$$= 0.75 \text{ m s}^{-2}$$

Force $F = ma = 800 \times 0.75 = 600$ N

Newton's first law of motion

Objects either stay at rest or move with constant velocity unless acted on by a force.

An object moving at constant velocity is either acted on by no forces, or the forces acting on it are balanced (so the resultant force is zero).

Newton's second law of motion for constant mass

When a resultant force F acts on an object of mass m, the object undergoes acceleration a such that

$$F = ma$$

Weight

The acceleration of a falling object acted on by gravity alone is equal to g. Because the force of gravity on the object is the only force acting on it, its **weight** (in newtons) $W = mg$, where m = the mass of the object (in kg).

- When an object is in equilbrium, the support force on it is equal and opposite to its weight.

- g is also referred to as the **gravitational field strength** at a given position. The gravitational field strength at the Earth's surface is 9.81 N kg^{-1}.

Inertia

The mass of an object is a measure of its **inertia**, which is its resistance to change of motion. More force is needed to give an object a certain acceleration than to give an object with less mass the same acceleration.

Summary questions

$g = 9.81$ N kg^{-1}

1. A heavy goods vehicle of mass 26 000 kg accelerates uniformly along a straight line from rest to a speed of 12 m s^{-1} in 50 s. Calculate:
   a. the acceleration of the vehicle (1 mark)
   b. the resultant force on the vehicle (1 mark)
   c. the ratio of the accelerating force to the weight of the car. (1 mark)

2. An aeroplane of mass 4000 kg lands on a runway at a speed of 62 m s^{-1} and stops in a distance of 1600 m. Calculate:
   a. the deceleration of the aeroplane (2 marks)
   b. the braking force on the aeroplane. (1 mark)

3. A golf ball of mass 0.027 kg was struck by a golf club with a force of 6.0 kN on level ground. The ball hit the ground 4.2 seconds later at a point 310 m away from the golfer.
   a. Calculate the acceleration of the ball when it was struck. (1 mark)
   b. Estimate:
      i. the average speed of the golf ball (1 mark)
      ii. the time of contact between the ball and the club. (2 marks)
   c. Discuss two assumptions made in your estimate in **b ii**. (4 marks)

8.2 Using *F= ma*

Specification reference: 3.4.1.5

When an object of mass m is acted on by two unequal forces F_1 and F_2 acting in opposite directions and F_1 is greater than F_2,

resultant force $F_1 - F_2 = ma$

where a is its acceleration, which is in the same direction as F_1.

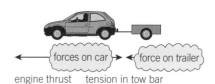
▲ **Figure 1** *Car and trailer*

1 A car pulling a trailer

Consider the example of a car of mass M fitted with a trailer of mass m on a level road. When the car and the trailer accelerate, the car pulls the trailer forward and the trailer holds the car back. Assume air resistance is negligible.

If F is the driving force on the car pushing it forwards (from its engine thrust) and T is the tension in the tow bar, the resultant force:

- on the car = $F - T = Ma$

- on the trailer = $T = ma$ (because the tension T in the tow bar is the force pulling the trailer forwards).

Combining the two equations gives the driving force $F = Ma + ma = (M + m)a$

2 Lift forces

The resultant force on the lift = $T - mg$, where T is the tension in the lift cable and m is the total mass of the lift and occupants (Figure 2). Therefore $T - mg = ma$, where a = acceleration.

Table 1 shows how the tension in the cable depends on the motion of the lift.

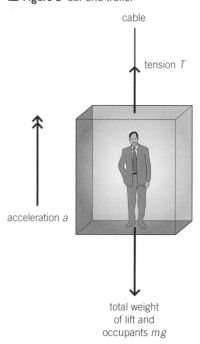
▲ **Figure 2** *In a lift*

Worked example

$g = 9.81\,\mathrm{m\,s^{-2}}$

A lift of total mass 650 kg starts moving downwards with an acceleration of $1.2\,\mathrm{m\,s^{-2}}$ for a brief time. Calculate the tension in the lift cable during this time.

Solution

The lift is moving down so its velocity $v < 0$. Since it accelerates, its acceleration a is in the same direction as its velocity, so $a < 0$.

Therefore, inserting $a = -1.2\,\mathrm{m\,s^{-2}}$ in the equation $T - mg = ma$ gives

$T = mg + ma = (650 \times 9.81) + (650 \times -1.2) = 5600\,\mathrm{N}$

▼ **Table 1**

Motion	Tension
constant velocity	$=mg$
upwards and accelerating	$>mg$
upwards and decelerating	$<mg$
downwards and decelerating	$>mg$
downwards and accelerating	$<mg$

Summary questions

$g = 9.81\,\mathrm{m\,s^{-2}}$

1 A rocket of mass 1050 kg blasts vertically from the launch pad at an acceleration of $4.80\,\mathrm{m\,s^{-2}}$. Calculate:
   a the weight of the rocket *(1 mark)*
   b the thrust of the rocket engines. *(2 marks)*

2 A lift and its occupants have a total mass of 1200 kg. Calculate the tension in the lift cable when the lift is:
   a descending at constant velocity *(1 mark)*
   b descending at a constant acceleration of $0.50\,\mathrm{m\,s^{-2}}$. *(3 marks)*

3 A skateboarder of mass 36 kg on a slope at $20°$ to the horizontal accelerates 5.0 m down the slope from rest in 3.0 s. Calculate:
   a the acceleration of the skateboarder *(2 marks)*
   b i the component of the skateboarder's weight acting down the slope
     ii the frictional force on the skateboarder. *(3 marks)*

Revision tip

When you use the equations in the examples opposite, make sure the two forces are the ones that act on the object under consideration not the forces that act on other objects. It's easy to get the forces mixed up when two or more objects interact.

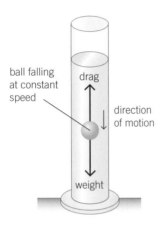

▲ **Figure 1** *Terminal speed*

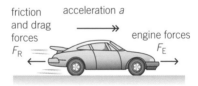

▲ **Figure 2** *Vehicle power*

Motion of an object falling in a fluid

Any object moving through a fluid experiences a force that drags on it due to the fluid. The **drag force** depends on:

- the shape of the object
- its speed and increases with increase of speed
- the viscosity of the fluid, which is a measure of how easily the fluid flows past a surface.

For an object released from rest in a fluid, its speed increases as it falls, so the drag force on it increases. The resultant force on the object is the difference between the force of gravity on it (its weight) and the drag force. As the drag force increases, the resultant force decreases, so the acceleration becomes less. If it continues falling, it attains **terminal speed**. (Figure 1).

At any instant, the resultant force $F = mg - D$, where m is the mass of the object and D is the drag force. Therefore, the acceleration of the object = $\frac{mg - D}{m} = g - \frac{D}{m}$.

The initial acceleration = g because the speed is zero, and therefore the drag force is zero, at the instant the object is released.

Motion of a powered vehicle

For a powered vehicle of mass m moving on a level surface, if F_E represents the **motive force** (driving force) provided by the engine, the resultant force on it = $F_E - F_R$, where F_R is the resistive force opposing the motion of the vehicle. (F_R = the sum of the drag forces acting on the vehicle.)

Therefore, its acceleration $a = \frac{F_E - F_R}{m}$

The maximum speed (the terminal speed) of the vehicle v_{max} is reached when the resistive force becomes equal and opposite to the engine force, and $a = 0$.

The top speed of a road vehicle or an aircraft depends on its engine power and its shape. A vehicle with a streamlined shape can reach a higher top speed than a vehicle with the same engine power that is not streamlined.

Summary questions

$g = 9.81\,\text{m s}^{-2}$

1 A metal ball of mass 0.045 kg, released from rest in a liquid, falls a distance of 0.32 m in 7.1 s. Assuming the ball reaches terminal speed within a fraction of a second, calculate:

   a its terminal speed   (*1 mark*)

   b the drag force on it when it falls at terminal speed.   (*2 marks*)

2 A steel ball fell vertically after it was released from rest above the surface of a liquid. The ball decelerated when it entered the liquid until its speed became constant.

   a Sketch a graph to show how the speed of the ball changed with time after it was released until it reached terminal speed. On your graph, label the point where the ball entered the water with the letter W.   (*2 marks*)

   b Explain why the ball decelerated after it entered the water.   (*5 marks*)

3 A vehicle of mass 22 000 kg has an engine which has a maximum engine force of 3800 N and a top speed of 36 m s^{-1} on a level road. Calculate:

   a its maximum acceleration from rest   (*1 mark*)

   b the total resistive force acting on it at top speed.   (*1 mark*)

8.4 On the road

Stopping distances

Thinking distance is the distance travelled by a vehicle in the time it takes the driver to react. For a vehicle moving at constant speed v, the thinking distance s_1 = speed × reaction time = vt_0, where t_0 is the reaction time of the driver.

The reaction time of a driver is affected by distractions, drugs, and alcohol.

Braking distance is the distance travelled by a car in the time it takes to stop safely, from when the brakes are first applied. Assuming constant deceleration, a, to zero speed from speed u, the braking distance $s_2 = \dfrac{u^2}{2a}$ since $u^2 = 2as_2$.

The braking distance for a vehicle depends on the vehicle speed before the brakes are applied, on the road conditions, and on the condition of the tyres.

Stopping distance = thinking distance + braking distance = $s_1 + s_2$

> **Revision tip**
>
> Remember the values for decelerations are negative as a deceleration is in the opposite direction to the velocity.

Summary questions

$g = 9.81\,\mathrm{m\,s^{-2}}$

1 A vehicle is travelling at a speed of $20\,\mathrm{m\,s^{-1}}$ on a level road, when the driver sees a pedestrian stepping off the pavement into the road 58.0 m ahead. The driver reacts within 0.70 s and applies the brakes, causing the car to decelerate at $5.1\,\mathrm{m\,s^{-2}}$ and stop safely a short distance from the pedestrian.
   a Calculate: **i** the thinking distance **ii** the braking distance. (*3 marks*)
   b How far does the driver stop from where the pedestrian stepped into the road? (*1 mark*)

2 A vehicle travelling at a speed of $28\,\mathrm{m\,s^{-1}}$ on a dry level road brakes to a standstill in a distance of 71 m. Calculate:
   a the deceleration of the vehicle from this speed to a standstill over this distance (*2 marks*)
   b the braking force on a vehicle of mass 15000 kg on this road as it stops. (*1 mark*)

3 a Explain why the stopping distance on a wet road is longer than the stopping distance at the same speed on the same road when it is dry. (*2 marks*)
   b The braking force on a vehicle travelling on a certain type of level road surface is $0.57 ×$ the vehicle's weight. For a vehicle of mass 2400 kg, calculate the braking distance on this road for a speed of $31\,\mathrm{m\,s^{-1}}$. (*4 marks*)

Worked example

$g = 9.81\,\mathrm{m\,s^{-2}}$

A vehicle of mass 1200 kg, travelling on a level road at a speed of $14\,\mathrm{m\,s^{-1}}$, is brought to a standstill without skidding in a distance of 25.0 m. Calculate:

a the deceleration of the vehicle

b the braking force.

Solution

a $u = 14\,\mathrm{m\,s^{-1}}, v = 0, s = 25.0\,\mathrm{m}$

   To calculate a, rearrange $v^2 = u^2 + 2as$ to give $a = \dfrac{v^2 - u^2}{2s} = \dfrac{0 - 14^2}{2 \times 25} = $ 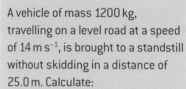 $-3.92\,\mathrm{m\,s^{-2}}$

   Therefore, the deceleration is $3.92\,\mathrm{m\,s^{-2}}$

b Braking force = mass × acceleration = $1200 \times 3.92 = 4700\,\mathrm{N}$

8.5 Vehicle safety

Specification reference: 3.4.1.5

Impact force

For a collision in which the velocity of an object of mass m changes from initial velocity u to final velocity v in a distance s, and

- the **impact time** $t = \dfrac{2s}{u + v}$ the acceleration $a = \dfrac{v - u}{t}$

- the **impact force** $F = ma$

Contact time and impact time

When objects collide and bounce off each other, they are in contact with each other for a certain time which is the same for both objects. The impact time is equal to the contact time in this case. When two vehicles collide and do *not* separate from each other after the collision, the impact time is not the same as the contact time.

Car safety features

The following vehicle safety features are designed to increase the impact time and so reduce the impact force.

- *Vehicle bumpers* give way a little in a low-speed impact and so increase the impact time.

- *Crumple zones* are designed to crumple in a front-end impact and so increase the impact time for the engine compartments.

- *Seat belts* restrain their wearer from crashing into the vehicle frame after the vehicle suddenly stops. The seat belt brings the wearer to a stop more gradually than without it.

- *Collapsible steering wheels* are designed to collapse in a front-end impact, if the driver makes contact with the steering wheel, thus increasing the impact time.

- *Airbags* reduce the force on a person, because they act as a cushion and increase the impact time on the person. Also, an airbag spreads the impact force over a wider area of the body so it has less effect on the body.

Synoptic link

The impact force can also be worked out using the equation

$F = \dfrac{\text{change of kinetic energy}}{\text{Impact distance}}$

See Topic 10.2, Kinetic energy and potential energy.

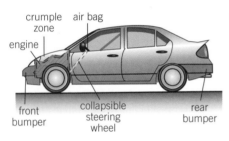

▲ **Figure 1** *Vehicle safety features*

Summary questions

$g = 9.81\ \mathrm{m\,s^{-2}}$

1   A car of mass 1100 kg travelling at a speed of 12 m s^{-1} is struck from behind by another vehicle, causing its speed to increase to 18 m s^{-1} in a distance of 3.0 m. Calculate the impact force on the car.   *(3 marks)*

2   The front bumper of a car of mass 900 kg is capable of withstanding an impact with a stationary object, provided the car is not moving faster than 3.0 m s^{-1}, when the impact occurs. The impact time at this speed is 0.40 s. Calculate the impact force on the car.   *(2 marks)*

3   In a crash, a vehicle travelling at a speed of 23 m s^{-1} stops after 4.8 m. A passenger of mass 62 kg is wearing a seat belt, which restrains her forward movement relative to the car to a distance of 0.5 m.
    a   Calculate the resultant force on the passenger.   *(3 marks)*
    b   Explain why the force on the passenger would have been much larger if she had not been wearing a seat belt.   *(3 marks)*

1 A vehicle of mass 1500 kg towing a trailer of mass 300 kg on a level road accelerates from rest to a velocity of 8.8 m s⁻¹ in 40 s, without change of direction.

    a   **i**   Calculate the force that accelerated the vehicle and the trailer.

         ii Calculate the tension in the tow bar during this time. Assume the tow bar is horizontal   *(3 marks)*

    b Calculate how long the vehicle without the trailer would take to accelerate with the same force from rest to a velocity of 8.8 m s⁻¹.   *(2 marks)*

2 A cyclist on a straight level section of a road accelerated from rest at a constant acceleration and reached a speed of 8.3 m s⁻¹ in a distance of 74 m. The total mass of the cyclist and the cycle was 61 kg.

    a Calculate the resultant force on the cyclist and the cycle.   *(3 marks)*

    b The cyclist then used the same force to travel on a downhill section which was inclined at an angle of 5.0° to the horizontal. Calculate the acceleration on this section.   *(4 marks)*

3 Figure 1 shows how the velocity of a motor car increased with time as it accelerated from rest along a straight horizontal road.

    a Use the graph to determine the maximum acceleration of the car.   *(2 marks)*

    b Throughout the motion shown in Figure 1 the driving force acting on the car was reduced gradually.

         i   The mass of the car and its contents was 1200 kg. Calculate the initial driving force.

         ii When the car was travelling at constant velocity, the driving force was 650 N. What was the magnitude of the resistive force acting on the car at this velocity? Give a reason for your answer.   *(7 marks)*

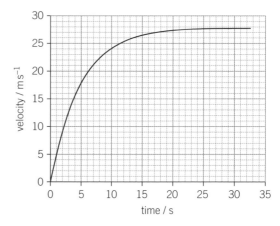

▲ **Figure 1**

4 A fairground ride ends with a car moving up a ramp of length 28 m at a slope of 32° to the horizontal then onto a flat section where it brakes and stops.

▲ **Figure 2**

    a The car enters at the bottom of the ramp at a speed of 18 m s⁻¹ and leaves it at 0.5 m s⁻¹.

         i   Calculate the deceleration of the car on the ramp.

         ii The mass of the car and its passengers is 750 kg. Calculate the resultant force on the car.   *(3 marks)*

    b **i**   Calculate the component of the weight of the car and the passengers parallel to the ramp when the car is on the ramp.

         ii Calculate the average resistive force on the car as it travels up the ramp.   *(4 marks)*

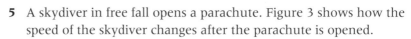

5 A skydiver in free fall opens a parachute. Figure 3 shows how the speed of the skydiver changes after the parachute is opened.

a

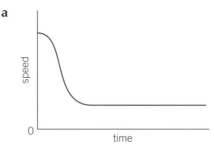

▲ Figure 3

Describe and explain how the acceleration changes with time. (*4 marks*)

b **i** Sketch a graph on a copy of Figure 3 to show how the speed would have changed if the weight of the skydiver had been less. Assume the parachute is opened at the same speed.

ii Explain why your graph differs from the one shown in Figure 3.

(*4 marks*)

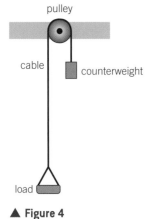

▲ Figure 4

6 A load of weight 250 N is lifted vertically using an overhead pulley and a cable attached to a counterweight of weight 290 N as shown in Figure 4.

a Assuming friction and air resistance are negligible:

i calculate the acceleration of the load

ii show that the tension in the cable is about 270 N. (*6 marks*)

b The total weight of the pulley and the cable is 55 N.
Calculate the support force on the pulley. (*2 marks*)

7 **a** In a car safety test, the front end of a certain type of car of mass 1800 kg travelling at a speed of $30 \, \text{m s}^{-1}$ crumpled in a distance of 0.80 m when the car hit a wall. Calculate:

i the impact time

ii the impact force. (*3 marks*)

b A test dummy in the front passenger seat was fitted with a seat belt which allowed the dummy to move a distance of 0.40 m before it stopped the dummy. Calculate the force exerted by the seat belt during the crash on the dummy. (*3 marks*)

8 A student on a skateboard accelerated from rest down a slope through a distance of 15 m in 5.6 s to the bottom of the slope then travelled onto flat level ground and gradually stopped.

a **i** Calculate the acceleration of the student down the slope.

ii The mass of the student was 38 kg. Calculate the resultant force on the student over this distance. (*3 marks*)

▲ Figure 5

b The slope was at an angle of 6.1° to the horizontal.

i Calculate the component of the student's weight down the slope.

ii Explain why the resultant force calculated in **a ii** was less than the component of the student's weight down the slope. (*4 marks*)

9.1 Momentum and impulse
9.2 Impact forces
Specification reference: 3.4.1.6

Momentum and Newton's laws of motion

For an object of mass m moving at velocity v, its momentum $p = mv$.

Newton's first law of motion: An object remains at rest or in uniform motion unless acted on by a force.

Newton's second law of motion: The rate of change of momentum of an object is proportional to the resultant force on it.

For a change of momentum $\Delta(mv)$ in time Δt, Newton's second law can be written as

$$F = \frac{\Delta(mv)}{\Delta t}$$

where the unit of force (the **newton**) is defined as the amount of force that gives an object of mass 1 kg an acceleration of $1\,\text{m\,s}^{-2}$.

1. If mass is constant, then $\Delta(mv) = m\Delta v$, where Δv is the change of velocity of the object.
   $\therefore F = \dfrac{m\Delta v}{\Delta t} = ma$ where acceleration $a = \dfrac{\Delta v}{\Delta t} \left(= \dfrac{v-u}{t}$ for constant acceleration$\right)$

2. If mass is lost or gained at constant velocity v, then $\Delta(mv) = v\Delta m$, where Δm is the change of mass of the object.
   $\therefore F = v\dfrac{\Delta m}{\Delta t}$ where $\dfrac{\Delta m}{\Delta t}$ = change of mass per second

This form of Newton's second law is used in any situation where an object gains or loses mass continuously (e.g., water from a hose pipe or hot gas from a jet engine).

Worked example

A pump forced water out of a hose pipe at a rate of 6.6 kilograms per minute with a velocity of $19\,\text{m\,s}^{-1}$. Calculate the force on the hose pipe due to the water.

Solution

Mass of water lost per second $= \dfrac{6.6\,\text{kg}}{60\,\text{s}} = 0.11\,\text{kg\,s}^{-1}$

Force $= v\dfrac{\Delta m}{\Delta t} = 19 \times 0.11 = 2.1\,\text{N}$

$$\text{impulse} = F\Delta t = \Delta(mv)$$

Hence, the impulse of a force acting on an object is equal to the change of momentum of the object.

Force–time graphs

A force–time graph can be used to find the change of momentum of an object. This is because the area under the line represents the impulse of the resultant force on the object and therefore gives the change of momentum. See Figure 1.

The area under the line of a force–time graph represents the change of momentum or the impulse of the force.

Revision tip

Momentum is a vector quantity. Its direction is the same as the direction of the object's velocity.

Revision tip

The unit of momentum may be either $\text{kg\,m\,s}^{-1}$ or (more neatly) N s.

Key term

The **impulse** of a force is defined as the force × the time for which the force acts.

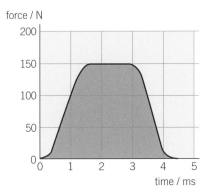

The area under the line is approximately 9 blocks. Each block represents an impulse of 0.050 N s (= 50 N × 1 ms). So the total impulse and therefore the change of momentum is approximately 0.45 N s

▲ **Figure 1** *Force against time for an impact*

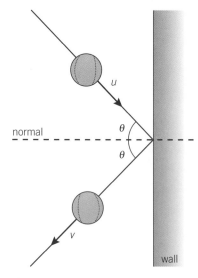

▲ **Figure 2** *An oblique impact*

Rebound impacts

Consider a ball that hits a wall at speed u in an oblique impact and rebounds at the same speed and the same angle to the wall.

The normal component of its momentum is $+mu \cos \theta$ before the impact and $-mu \cos \theta$ after the impact.

Therefore, its change of momentum $= (-mu \cos \theta) - (mu \cos \theta) = -2mu \cos \theta$

Note: If $\theta = 0$, the initial velocity is perpendicular (i.e., normal) to the wall so the change of momentum $= -2mu$.

Summary questions

1 An aircraft of total mass 25 000 kg accelerated on a runway from rest to a velocity of 160 m s^{-1} after 58 s when it took off. Calculate the engine force during this time. *(2 marks)*

2 The velocity of a vehicle of mass 1300 kg was reduced from 20 m s^{-1} by a constant force of 1800 N which acted for 4.0 s, then by a force that decreased uniformly to zero in a further 6.0 s.
   a Sketch the force–time graph for this situation. *(1 mark)*
   b Use the force–time graph to determine the total change of momentum and hence show that the final velocity of the vehicle was about 10 m s^{-1}. *(3 marks)*

3 A van of mass 1500 kg travelling at a speed of 2.0 m s^{-1} was struck from behind by another vehicle. The impact lasts for 0.55 s and causes the speed of the van to increase to 6.5 m s^{-1}. Calculate the impact force. *(2 marks)*

4 A molecule of mass 5.0×10^{-26} kg moving at a speed of 420 m s^{-1} hits a surface at 60° to the normal and rebounds without loss of speed at 60° to the normal in an impact lasting 0.22 ns. Calculate the force on the molecule. *(3 marks)*

9.3 Conservation of momentum
9.4 Elastic and inelastic collisions
9.5 Explosions

Specification reference: 3.4.1.6

Newton's third law of motion

When two objects interact, they exert equal and opposite forces on each other.

In other words, if object A exerts a force on object B, there must be an equal and opposite force acting on object A due to object B.

The principle of conservation of momentum

For any system of interacting objects, their total momentum remains constant, provided no external resultant force acts on the system.

Interactions between objects can transfer momentum between them. But their total momentum does not change. Figure 1 shows two snooker balls A and B of masses m_A and m_B that move along a straight line, before and after they collide.

The total initial momentum $= m_A u_A + m_B u_B$

The total final momentum $= m_A v_A + m_B v_B$

Because the total momentum is conserved,

$$m_A v_A + m_B v_B = m_A u_A + m_B u_B$$

If the colliding objects stick together as a result of the collision, they have the same final velocity. The above equation with V as the final velocity may therefore be written

$$(m_A + m_B)V = m_A u_A + m_B u_B$$

Elastic and inelastic collisions

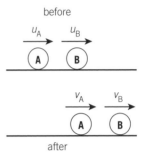
▲ Figure 1

Revision tip

The vector nature of momentum needs to be taken into account in straight-line collisions by defining one direction as + and the opposite direction as −.

Key terms

An **elastic collision** is one where there is no loss of kinetic energy.

An **inelastic collision** occurs where the colliding objects have less kinetic energy after the collision than before the collision.

Synoptic link

The kinetic energy of an object can be worked out using the kinetic energy equation $E_K = \frac{1}{2}mv^2$, where m is the mass of the object and v is its speed (see Topic 10.2, Kinetic energy and potential energy).

Worked example

A railway wagon X of mass 8000 kg moving at $3.0\,\text{m s}^{-1}$ collides with wagon Y of mass 5000 kg moving towards X at a speed of $0.5\,\text{m s}^{-1}$. After the collision, the two wagons separate and X moves at a speed of $1.0\,\text{m s}^{-1}$ without change of direction. Calculate:

a the speed and direction of Y after the collision

b the loss of kinetic energy due to the collision.

Solution

a Let the initial direction of X be the + direction so the velocity of Y before the collision $= -0.5\,\text{m s}^{-1}$.

The total initial momentum $= (8000 \times 3.0) + (5000 \times -0.5) = 21\,500\,\text{kg m s}^{-1}$.

The total final momentum $= (8000 \times 1.0) + 5000V$, where V is the speed of the 5000 kg wagon after the collision.

Using the principle of conservation of momentum

$8000 + 5000V = 21\,500$

$5000V = 21\,500 - 8000 = 13\,500$

$V = \dfrac{13\,500}{5000} = 2.7\,\text{m s}^{-1}$

b Kinetic energy of X E_{Kx} before the collision $= \frac{1}{2} \times 8000 \times 3.0^2 = 36\,000$ J

E_{Ky} before the collision $= \frac{1}{2} \times 5000 \times 0.5^2 = 625$ J

E_{Kx} after the collision $= \frac{1}{2} \times 8000 \times 1.0^2 = 4000$ J

E_{Ky} after the collision $= \frac{1}{2} \times 5000 \times 2.7^2 = 18\,225$ J

$\therefore$ loss of kinetic energy due to the collision $= (36\,000 + 625) - (4000 + 18\,225)$
$= 14\,400$ J

Explosions

When two objects A and B of masses m_A and m_B fly apart after being initially at rest, they recoil from each other with equal and opposite amounts of momentum.

Using the principle of the conservation of momentum, their total momentum immediately after the explosion $= 0$ as the total initial momentum is zero.

Therefore, $m_A v_A + m_B v_B = 0$ where v_A and v_B are their velocities when they move apart.

So, $m_B v_B = -m_A v_A$, where the minus sign means that the two masses move apart in opposite directions.

Summary questions

1 In a laboratory experiment, trolley A of mass 1.50 kg moving at a speed of $0.35\,\text{m s}^{-1}$ collides with trolley B moving in the opposite direction at a speed of $0.25\,\text{m s}^{-1}$. The two trolleys couple together on collision and move in the initial direction of A at a speed of $0.050\,\text{m s}^{-1}$ immediately after the collision. Calculate the mass of B. (4 marks)

2 A ball X of mass 0.60 kg moving at a speed of $3.0\,\text{m s}^{-1}$ along a straight line collides with a ball Y of mass 0.40 kg which was initially stationary. As a result of the collision, X has a velocity of $0.80\,\text{m s}^{-1}$ in the same direction along the line. Calculate the speed and direction of Y immediately after the collision. (4 marks)

3 A rail wagon P of mass 3000 kg moving at a velocity of $1.2\,\text{m s}^{-1}$ collides with a wagon Q of mass 2000 kg moving at a speed of $2.5\,\text{m s}^{-1}$ in the opposite direction. After the collision, Q moves at a velocity of $0.50\,\text{m s}^{-1}$ in the direction P was originally moving in.
   a Calculate the speed and direction of P after the collision. (4 marks)
   b Show that the collision is inelastic. (3 marks)

4 A shell of mass 2.4 kg is fired at a speed of $150\,\text{m s}^{-1}$ from an artillery gun of mass 900 kg.
   a Calculate the recoil velocity of the gun. (2 marks)
   b Calculate the kinetic energy of the gun as a percentage of the total kinetic energy of the shell and the gun. (2 marks)

1 **a** State the principle of conservation of momentum. *(1 mark)*

b A vehicle of mass 1200 kg moving at a speed of 21.0 m s^{-1} collided with a vehicle of mass 1600 kg moving in the same direction at a speed of 3.0 m s^{-1}. The two vehicles locked together on impact. Calculate:

   i the velocity of the two vehicles immediately after impact

   ii the loss of kinetic energy due to the impact. *(6 marks)*

2 A football of mass 0.44 kg travelling at a speed of 24 m s^{-1} strikes a wall at an angle of 30° to the normal and rebounds at the same speed at the same angle to the normal.

a Calculate the change of momentum of the ball *(1 mark)*

b The ball was in contact with the wall for 95 ms. Calculate the impact force on the wall. *(2 marks)*

3 **a** State what is meant by an elastic collision. *(1 mark)*

b A ball X of mass 0.12 kg moving at a speed of 0.58 m s^{-1} collides head-on with a second ball Y of mass 0.10 kg moving in the opposite direction at a speed of 0.50 m s^{-1}. After the impact, X continues in the same direction at a speed of 0.15 m s^{-1}.

   i Calculate the speed and direction of Y after the collision. *(3 marks)*

   ii Show that the collision was inelastic. *(3 marks)*

4 Two trolleys, A of mass 1.20 kg and B of mass 0.80 kg, are initially stationary on a level track.

a When a trigger is pressed on one of the trolleys, a spring pushes the two trolleys apart and B moves away at a velocity of 0.15 m s^{-1}.

   i Calculate the velocity of A.

   ii Calculate the total kinetic energy of the two trolleys immediately after the explosion. *(6 marks)*

b In part **a**, if the test had been carried out with A held firmly, calculate the speed at which B would have recoiled, assuming the energy stored in the spring before release is equal to the total kinetic energy calculated in **a ii**. *(2 marks)*

5 **a** When an α particle is emitted from a nucleus of the bismuth isotope, a nucleus of thallium (Th) is formed. Complete the equation below.

$$^{210}_{83}\text{Bi} \rightarrow \alpha + \text{Th}$$

(2 marks)

b The α particle in part A is emitted at a speed of 1.5×10^7 m s^{-1}.

   i Calculate the speed of recoil of the thallium nucleus immediately after the α particle has been emitted. Assume the parent nucleus is initially at rest.

   ii The mass of the α particle is 6.7×10^{-27} kg. Calculate the kinetic energy of the α particle immediately after it has been emitted as a percentage of the energy released by the bismuth nucleus. Ignore relativistic effects. *(6 marks)*

6 Figure 1 shows how the force, F, on a ball varied with time, t, when a tennis racquet was used to hit the ball horizontally when the ball was momentarily stationary.

a State what the area under the curve represents. *(1 mark)*

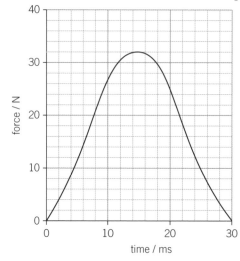

▲ **Figure 1**

b i Use the graph to determine the change of momentum of the ball.

ii The mass of the ball was 0.056 kg. Calculate the speed at which the ball left the racquet.

iii Estimate the average acceleration of the ball during the impact. *(6 marks)*

7 Figure 2 shows a pile driver used to hammer a steel pile vertically into the ground. The pile driver hammer is repeatedly raised above the top end of the pile then released so it drops onto the top end of the pile and hammers the pile deep into the ground.

a The hammer has a mass of 4000 kg and is raised to a height of 0.80 m above the top end of the pile then released.

i Calculate the speed and the momentum of the hammer just before it makes contact with the pile.

ii The pile has a mass of 2000 kg. Calculate the velocity of the hammer and the pile immediately after the impact. *(6 marks)*

b Each time the hammer is dropped onto the pile, the pile is pushed 20 mm further into the ground.

i Calculate the deceleration of the pile as it is pushed into the ground.

ii Calculate the average force of friction on the pile as it is pushed into the ground. *(3 marks)*

8 A water jet cutter used in industry to cut materials in fine detail produces a jet of water at a speed of 400 m s^{-1} and a flow rate of 2.7 kg per minute.

a Calculate the momentum transferred per second by the water. *(3 marks)*

b Calculate the force exerted by the water jet when it is directed normally onto a surface. Assuming the water does not rebound from the surface. *(2 marks)*

c The water jet emerges from a nozzle of diameter 0.38 mm. Estimate the pressure of the water jet at normal incidence on the surface. (Hint: Pressure = force per unit area). *(2 marks)*

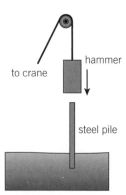

▲ **Figure 2**

10.1 Work and energy
10.2 Kinetic energy and potential energy
Specification reference: 3.4.1.7; 3.4.1.8

Work is done on an object when a force acting on it makes it move in the direction of the force. The work done is calculated using the equation

work done = force × distance moved in the direction of the force

The unit of work, the joule (J), is the work done when a force of 1 N moves its point of application by a distance of 1 m in the direction of the force.

When a force F acts on an object and moves it through a distance s along a straight line at an angle to the force direction, the work done W by the force is given by $W = Fs \cos \theta$.

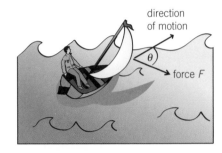

▲ **Figure 1** *Force and displacement*

Force–distance graphs

The area under the line of a force–distance graph represents the total work done.

Figure 2 shows a force–distance graph for a variable force that moves an object in the direction of the force. The work done to move the object through any distance can be determined from the area under the part of the line corresponding to that distance.

Figure 3 shows how the force needed to extend a spring varies with the extension of the spring. The graph is a straight line through the origin provided the spring obeys Hooke's Law. By considering the area under the line (which is a triangle), it can be shown that the work done and therefore

the energy stored in the spring $= \frac{1}{2} F \Delta L$

Revision tip

No work is done when F and s are at right angles to each other.

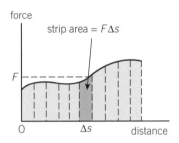

▲ **Figure 2** *Force–distance graph for a variable force*

Energy

Energy is measured in joules. When a force does work, energy is transferred by the force equal to the work done.

The **kinetic energy** of an object of mass m moving at speed v is given by the equation

kinetic energy, $E_K = \frac{1}{2} mv^2$

The change of **gravitational potential energy** ΔE_P of an object of mass m raised through a vertical height Δh, is given by the equation

$$\Delta E_P = mg \Delta h$$

At the Earth's surface, $g = 9.81 \, \text{m s}^{-2}$.

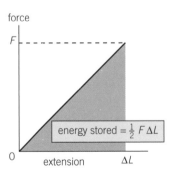

▲ **Figure 3** *Force against extension for a spring*

Conservation of energy

Whenever energy is transferred, the total amount of energy after the transfer is always equal to the total amount of energy before the transfer. The total amount of energy is unchanged.

Energy cannot be created or destroyed.

This statement is known as the **principle of conservation of energy**.

Synoptic link

The force needed to stretch a spring is proportional to the extension of the spring. This is known as Hooke's law. For more about springs see Topic 11.2, Springs.

Energy changes involving kinetic and potential energy

Consider an object of mass m that is released from rest. Suppose its speed is v after it descends through a vertical distance Δh. If the resistive forces on it are negligible, its gain of kinetic energy is equal to its loss of potential energy. Therefore

$$\frac{1}{2}\,mv^2 = mg\,\Delta h$$

If the resistive forces are *not* negligible, its gain of kinetic energy is less than its loss of potential energy because work W is done by the resistive forces. In other words,

$$\frac{1}{2}mv^2 = mg\,\Delta h - W$$

Therefore, the work done by the resistive forces $W = mg\,\Delta h - \frac{1}{2}mv^2$.

Summary questions

$g = 9.81\,\mathrm{m\,s^{-2}}$

1 Calculate the work done when:
   a a force of 14 N moves an object by a distance of 3.0 m in a direction at 60° to the direction of the force *(1 mark)*
   b a spring that obeys Hooke's Law is stretched from zero extension to an extension of 0.45 m by a force that increases to 20 N. *(1 mark)*

2 A ball of mass 0.048 kg was thrown directly downwards at a speed of $16.0\,\mathrm{m\,s^{-1}}$ from the top of a tower of height 23.0 m. Calculate:
   a its kinetic energy *(2 marks)*
   b its speed just before it hit the ground. Assume air resistance is negligible. *(2 marks)*

3 A skier of mass 80 kg (including the skis) reaches a speed of $14\,\mathrm{m\,s^{-1}}$ after skiing from rest 560 m down a slope onto level ground which is 40 m lower than the starting point. Calculate:
   a the loss of potential energy *(1 mark)*
   b the gain of kinetic energy of the skier and the skis *(1 mark)*
   c the average resistive force during the descent. *(2 marks)*

10.3 Power
10.4 Energy and efficiency

Specification reference: 3.4.1.7

Power is defined as the rate of transfer of energy.

The unit of power is the watt (W), equal to an energy transfer rate of 1 joule per second. Note that 1 kilowatt (kW) = 1000 W, and 1 megawatt (MW) = 10^6 W.

If energy ΔE is transferred steadily in time Δt,

$$\text{power } P = \frac{\Delta E}{\Delta t}$$

Engine power

Vehicle engines, marine engines, and aircraft engines are all designed to make objects move. The output power of an engine is sometimes called its **motive power**.

1 For a powered vehicle driven by a constant force F moving at speed v, the output power of the engine $P = Fv$.

The work done by the engine is transferred into the internal energy of the surroundings by the resistive forces.

2 For a powered vehicle that gains speed when moving horizontally, the work done by the engine increases the kinetic energy of the vehicle and enables the vehicle to overcome the resistive forces acting on it. Therefore,

$$\begin{array}{ccc} \textbf{the output power} & \textbf{energy per second wasted} & \textbf{the gain of kinetic} \\ \textbf{of the engine} & = \textbf{due to the resistive force} & + \textbf{energy per second} \end{array}$$

> ### Worked example 🖩
>
> When an aircraft is in level flight at a constant velocity of 120 m s^{-1}, its engines have a total power output of 4.0 MW. Calculate the driving force of the engine at this speed.
>
> #### Solution
>
> Rearranging power = force × velocity gives the driving force,
>
> $$F = \frac{\text{output power}}{\text{speed}} = \frac{4.0 \times 10^6}{120} = 33 \text{ kN}$$

Machines at work

If a machine exerts a force F on an object to make it move at a constant velocity v,

$$\text{the output power of the machine } P_{OUT} = Fv$$

Efficiency measures

Useful energy is energy transferred for a purpose. In any machine, some of the energy supplied to it is wasted, usually as energy transferred to the surroundings by:

• sound waves created by vibrating machinery, or
• heating due to friction or air resistance.

$$\begin{aligned} \textbf{The efficiency of a machine} &= \frac{\textbf{useful energy transferred by the machine}}{\textbf{energy supplied to the machine}} \\ &= \frac{\textbf{work done by the machine}}{\textbf{energy supplied to the machine}} \end{aligned}$$

Synoptic link

The electrical power supplied to a component = IV where I is the current in the component when the pd across it is V. See Topic 12.2, Potential difference and power, for more about electrical power.

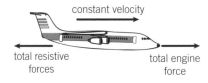

constant velocity

total resistive forces   total engine force

▲ **Figure 1** *Engine power*

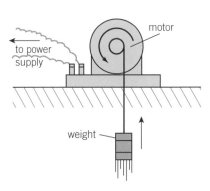

motor

to power supply

weight

▲ **Figure 2** *Efficiency*

Note:

- **Efficiency** can be expressed in terms of power as

$$\frac{\text{the output power of a machine}}{\text{the input to the machine}}$$

- *Percentage* efficiency = efficiency × 100%

Renewable energy

- Most of the energy we use at present is obtained from fossil fuels. Scientists think that the use of fossil fuels is causing climate change, due to the increasing amount of carbon dioxide in the atmosphere.

- Carbon emissions can be cut by building new nuclear power stations and developing more renewable energy resources.

- The output power of many renewable energy resources can be estimated using AS/A Level Physics principles and formulae. For example, the output power of a wind turbine is proportional to v^3 where v is the wind speed. This is because:

  - the kinetic energy of air moving at speed v is proportional to v^2

  - the mass of air passing through the area swept out by the blades is proportional to v

  - so the kinetic energy per second of the air passing through a wind turbine is proportional to v^3.

Summary questions

$g = 9.81\,\mathrm{m\,s^{-2}}$

1 A heavy goods vehicle of mass 40 000 kg moving at a constant velocity of 31 m s^{-1} has an output power of 240 kW.
   a Calculate the motive force of its engine at this speed. *(2 marks)*
   b The engine has an efficiency of 37% at this speed. Calculate the energy per second wasted at this speed. *(2 marks)*

2 The maximum power that can be obtained from a wind turbine is proportional to the cube of the wind speed. When the wind speed is 12 m s^{-1}, the power output of a certain wind turbine is 1.4 MW.
   a Calculate the power output of this wind turbine when the wind speed is 8.0 m s^{-1}. *(3 marks)*
   b The efficiency of the wind turbine at 12 m s^{-1} is 48%. Calculate the kinetic energy per second of the wind passing through the turbine. *(2 marks)*

3 A hydroelectric power station produces electrical power at an overall efficiency of 28%. The power station is driven by water from an upland reservoir 350 m above the power station. Calculate the output power when the water passes through it at a flow rate of 52 m^3 per second. The density of water is 1000 kg m^{-3}. *(3 marks)*

Chapter 10 Practice questions

g = 9.81 m s⁻²

1 A skateboarder of mass 52.0 kg at the top of a slope travels a distance of 10.2 m down the slope and reaches a speed of 2.12 m s⁻¹ at the bottom of the slope which is 2.50 m lower than the top.

   a Calculate the skateboarder's:

      i loss of potential energy

      ii gain of kinetic energy. *(2 marks)*

   b The average frictional force on the skateboarder during the descent. *(3 marks)*

2 A child of mass 35 kg on a trampoline bounces up and down repeatedly through a vertical distance of 0.90 m between the lowest and the highest position of her centre of mass.

   a Describe the energy changes that take place when she descends from her highest position to her lowest position. *(3 marks)*

   b Each time she descends from her highest position, she accelerates due to gravity then decelerates vertically through a distance of 0.16 m to her lowest position.

      i Calculate her speed just before she starts to decelerate.

      ii Estimate the maximum energy that could be transferred to the trampoline in each bounce. *(4 marks)*

   c The trampoline is fitted with 90 springs, each with a spring constant of 3500 N m⁻¹. When the child is at her lowest position, the average extension of each spring is 42 mm. Calculate the energy stored in the springs at maximum extension. *(2 marks)*

3 A road truck of mass 12 × 10³ kg travels on a straight uphill road at a constant velocity of 18 m s⁻¹.

   a Calculate the kinetic energy of the truck at this speed. *(1 mark)*

   b The gradient of the road is at an angle of 5.0° to the horizontal as shown in Figure 1.

      i Calculate the gain of potential energy of the truck each second.

      ii The output power of the truck engine at this speed is 228 kW. Calculate the output force of the engine.

      iii Show that the total resistive force on the truck is 2.4 kN. *(7 marks)*

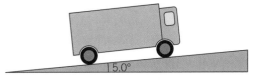

▲ **Figure 1**

4 **a** A tidal power station traps seawater over a flat area of 150 km² when the tide is 5.0 m above the power station turbines. The trapped water is released gradually over a period of 6 hours. Calculate:

      i the mass of trapped water

      ii the average loss of potential energy per second of this trapped water when it is released over a period of 6 hours

      The density of sea water = 1050 kg m⁻³. *(5 marks)*

   b A solar cell panel of area 1 m² can produces 220 W of electrical power on a sunny day. Calculate the area of panels that would generate electrical power at the same rate as the tidal power station in part **a**. *(1 mark)*

5 In the fairground ride shown in Figure 2, each carriage is pulled through a vertical height of 15 m to the top of an inclined track which then descends steeply to a level section which is partly in shallow water.

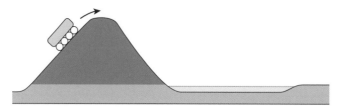

▲ **Figure 2**

a A carriage of total mass 1200 kg including its riders travels at a constant speed up the incline in 20 s, pulled by a conveyor belt driven by an electric motor operating at an efficiency of 58%. Estimate the minimum power supplied to the motor when it pulls the carriage up the track. *(3 marks)*

b The speed of the carriage at the start of its descent is 1.2 m s⁻¹. Calculate its speed when it reaches the level section of the track. *(4 marks)*

c The carriage leaves the shallow water at a speed of 11.5 m s⁻¹.

   i Calculate the energy transferred to the water.

   ii The length of track under water is 18.4 m. Estimate the average resistive force on the carriage when it passes through the water. *(4 marks)*

6 a A jet plane of mass 250 000 kg travels for 2400 m before it lifts off the runway at a speed of 150 m s⁻¹.

   i Calculate the kinetic energy of the aircraft at this speed.

   ii Estimate the average engine force during the time it travels along the runway. *(3 marks)*

b After taking off, the aircraft climbs to a height of 8500 m above the ground and travels at a constant velocity of 250 m s⁻¹.

   i Calculate the gain of potential energy of the plane in reaching this height from the ground.

   ii Calculate its kinetic energy when it moves at 250 m s⁻¹. *(2 marks)*

c Estimate the mass of fuel burned to take off from rest and reach the above height and speed. Assume the engines are 55% efficient and each kilogram of fuel used releases 30 MJ of energy. *(3 marks)*

7 A hydroelectric power station generates 450 MW of electrical power at an efficiency of 80%. The power station is driven by water that has descended from an upland reservoir 390 m above the power station.

a Calculate the volume of water per second passing through the turbines of the power station when its generators produce 450 MW of electrical power.

The density of water = 1000 kg m⁻³. *(6 marks)*

b The turbines at the power station are designed to pump water uphill to the reservoir using electrical power from electrical power generators elsewhere when the national demand for electrical power is low.

   i Estimate the percentage of the electrical power used for this purpose that is wasted. Assume the pumping process is also 80% efficient. *(3 marks)*

   ii Explain why the efficiency of this pumped storage process could never be more than 80%. *(3 marks)*

The **density of a substance** is defined as its mass per unit volume.

If volume V of a substance has mass m, its density ρ is given by

$$\rho = \frac{m}{V}$$

The unit of density is the kilogram per cubic metre ($kg\,m^{-3}$).

Rearranging the above equation gives $m = \rho V$ or $V = \frac{m}{\rho}$.

Density measurements

To measure the density of a substance, measure the mass and volume of a sample of the substance, and then calculate the density from $\frac{mass}{volume}$.

1 A regular solid

- Measure its mass using a top pan balance.

- Measure its dimensions using vernier callipers or a micrometer and calculate its volume using the appropriate equation. For example, for a sphere of radius r, volume $= \frac{4}{3}\pi r^3$ (see Figure 1 for other volume equations).

2 A liquid

- Measure the mass of an empty measuring cylinder. Pour some of the liquid into the measuring cylinder and measure the volume of the liquid directly. Use as much liquid as possible to reduce the percentage error in your measurement.

- Measure the mass of the cylinder and liquid to enable the mass of the liquid to be calculated.

3 An irregular solid

- Measure the mass of the object.

- Lower the object on a thread into the liquid in a measuring cylinder and measure the increase in the liquid level. This is the volume of the object.

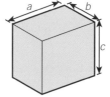

a *Volume of cuboid $= a \times b \times c$*

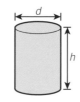

b *Volume of cylinder $= \frac{\pi d^2}{4} \times h$*

▲ **Figure 1** *Volume equations*

Density of alloys

An alloy is a solid mixture of two or more metals. For an alloy, of volume V and mass m, that consists of two metals A and B:

- if the volume of metal A $= V_A$, the mass of metal A $m_A = \rho_A V_A$, where ρ_A is the density of metal A

- if the volume of metal B $= V_B$, the mass of metal B $m_B = \rho_B V_B$, where ρ_B is the density of metal B.

Given the density of each metal and how much of each is in the alloy, these equations can be used to work out the density of the alloy.

Worked example

A brass object consists of 0.62 kg of copper and 0.29 kg of zinc. Calculate the volume and the density of this object. The density of copper $= 8900\,kg\,m^{-3}$. The density of zinc $= 7100\,kg\,m^{-3}$.

Solution

Volume of copper = mass of copper ÷ density of copper

$= 0.62\,kg \div 8900\,kg\,m^{-3} = 7.0 \times 10^{-5}\,m^3$

Volume of zinc = mass of zinc ÷ density of zinc

$= 0.29 \, kg \div 7100 \, kg \, m^{-3} = 4.1 \times 10^{-5} \, m^3$

Total mass $m = 0.62 + 0.29 = 0.91 \, kg$

Total volume $V = 1.1 \times 10^{-4} \, m^3$

Density of alloy $\rho = \dfrac{m}{v} = \dfrac{0.91 \, kg}{1.1 \times 10^{-4} \, m^3} = 8270 \, kg \, m^{-3}$

Summary questions

1 A concrete paving stone has dimensions 2.0 cm × 60 cm × 75 cm and a mass of 22.4 kg. Calculate:

   a its volume   *(1 mark)*

   b its density.   *(1 mark)*

2 An empty tin of diameter 50 mm and of height 90 mm has a mass of 85 g. It is filled with a liquid to within 5 mm of the top. Its total mass is then 248 g. Calculate: **a** the mass **b** the volume **c** the density of the liquid in the tin.   *(3 marks)*

3 A metal wire has a diameter of 0.220 ± 0.005 mm, a length of 2415 ± 5 mm, and a mass of 0.730 ± 0.005 g.

   a Calculate:

      i the volume of the wire   *(2 marks)*

      ii the density of the metal in kg m^{-3}.   *(1 mark)*

   b Show that the uncertainty in the density value is about 430 kg m^{-3}.   *(4 marks)*

11.2 Springs

Hooke's law

Hooke's law states that the force needed to stretch a spring is directly proportional to the extension of the spring from its natural length.

Hooke's law may be written as **Force $F = k\Delta L$**

where k is the spring constant (or the stiffness constant) and ΔL is the extension from its natural length L. The unit of k is $\mathrm{N\,m^{-1}}$.

- The graph of F against ΔL is a straight line of gradient k through the origin.
- If a spring is stretched beyond its **elastic limit**, it does not return to its original length when the applied force is removed.

Spring combinations

Springs in parallel

Figure 2 shows a weight supported by means of two springs P and Q with spring constants k_P and k_Q in parallel with each other. The extension, ΔL, of each spring is the same.

The effective spring constant $k = k_P + k_Q$.

Springs in series

Figure 3 shows a weight supported by means of two springs joined end-on. The tension in each spring is the same and is equal to the weight W.

- $\dfrac{\text{the extension in P}}{\text{the extension in Q}} = \dfrac{k_Q}{k_P}$
- The effective spring constant k is given by the equation $\dfrac{1}{k} = \dfrac{1}{k_P} + \dfrac{1}{k_Q}$.

Key term

The **elastic limit** is the maximum length to which a material can be stretched and still return to its original length once the applied force is removed.

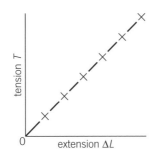

▲ **Figure 1** Hooke's law

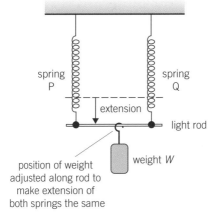

▲ **Figure 2** Two springs in parallel

position of weight adjusted along rod to make extension of both springs the same

---

Summary questions

$g = 9.81\,\mathrm{m\,s^{-2}}$

1. A steel spring has a spring constant of $40\,\mathrm{N\,m^{-1}}$. The spring hangs vertically and supports an $8.0\,\mathrm{N}$ weight at rest. Calculate:
   a. the extension of the spring *(1 mark)*
   b. the energy stored in the spring. *(1 mark)*

2. Two identical springs of length $250\,\mathrm{mm}$ are arranged in parallel and vertical as in Figure 2. When the springs support a $1.6\,\mathrm{N}$ weight, they both extend to a length of $274\,\mathrm{mm}$. Calculate:
   a. for each spring: **i** its tension *(1 mark)* **ii** its spring constant. *(3 marks)*
   b. the total energy stored in the springs. *(3 marks)*

3. Two unequal steel springs P and Q of length $250\,\mathrm{mm}$ are suspended vertically in series from a fixed point as in Figure 3. A $1.6\,\mathrm{N}$ weight is attached to the ends of the two springs. The lengths of the springs are then $280\,\mathrm{mm}$ for P and $300\,\mathrm{mm}$ for Q. Calculate:
   a. the tension in each spring *(1 mark)*
   b. the spring constant of each spring. *(4 marks)*

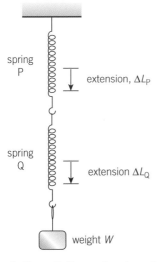

▲ **Figure 3** Two springs in series

11.3 Deformation of solids
11.4 More about stress and strain

Specification reference: 3.4.2.2

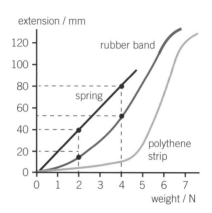

▲ **Figure 1** *Typical graphs*

Key terms

Elasticity is a property of a solid material that returns to its original shape after it has been deformed, once the forces that deformed it have been removed.

Tensile deformation occurs when an object is stretched.

Compressive deformation occurs when an object is compressed.

Tension–extension graphs

- A steel spring gives a straight line (see Figure 1), in accordance with Hooke's law (Topic 11.2, Springs).

- A rubber band at first extends easily when it is stretched then it becomes increasingly difficult to stretch further (see Figure 1).

- A polythene strip 'gives' and stretches easily after its initial stiffness is overcome then becomes more difficult to stretch (see Figure 1).

Tensile stress and tensile strain

- The **tensile stress** in the wire, $\sigma = \dfrac{T}{A}$, where T is the tension. The unit of stress is the **pascal** (Pa) equal to $1\,\text{Nm}^{-2}$.

- The **tensile strain** in the wire, $\varepsilon = \dfrac{\Delta L}{L}$, where ΔL is the extension (increase in length) of the wire. Strain is a ratio and therefore has no unit.

Figure 2 shows how the tensile stress in a wire varies with tensile strain.

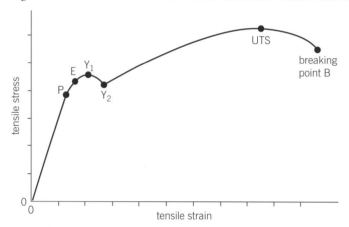

▲ **Figure 2** *Tensile stress versus tensile strain for a metal wire*

From zero to the limit of proportionality P, the tensile stress is proportional to the tensile strain. The value of $\dfrac{\text{stress}}{\text{strain}}$ is a constant, known as the **Young modulus** of the material.

$$\textbf{Young modulus } E = \frac{\text{tensile stress, } \sigma}{\text{tensile strain, } \varepsilon} = \frac{T}{A} \div \frac{\Delta L}{L} = \frac{TL}{A\Delta L}$$

- Beyond P, the **elastic limit** is the point beyond which the wire is permanently stretched and suffers **plastic deformation**.

- At the **yield point** Y_1, the wire weakens temporarily. Beyond Y_2, a small increase in the tensile stress causes a large increase in tensile strain as the material of the wire undergoes plastic flow.

- Beyond maximum tensile stress, the **ultimate tensile stress** (UTS) (or breaking stress), the wire loses its strength, extends, becomes narrower and breaks at point B.

Revision tip

A **brittle** material snaps without any noticeable yield.

A **ductile** material stretches considerably as the tension is increased before it breaks.

Loading and unloading curves

The area under a tension against extension line gives the work done. If the loading and unloading curves differ, energy transferred to the material is not all recovered as useful energy when the material unstretches.

1 For a metal wire, the loading and unloading lines are the same up to the elastic limit so all the energy stored in the wire can be recovered, provided the elastic limit is not exceeded.

2 For a rubber band, the unloading curve is below the loading curve except at zero and maximum extensions. The area between the curves gives the internal energy gained by the rubber band when it unstretches.

3 For a polythene strip, it does not return to its initial length when it is completely unloaded. The area between the loading and unloading curves represents work done to deform the material permanently, as well as internal energy retained by the polythene when it unstretches.

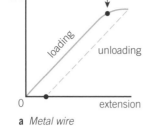

a *Metal wire*

b *Rubber*

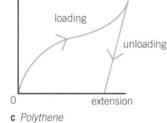

c *Polythene*

▲ **Figure 3** *Loading and unloading curves*

Worked example

A steel wire of uniform diameter 0.26 mm and of length 1540 mm is stretched to a tension of 58 N. Calculate the extension in the wire at this tension.

The Young modulus for steel = 2.1×10^{11} Pa.

Solution

Tension $T = 58$ N

Area of cross section of wire $= \frac{1}{4}\pi d^2 = \frac{1}{4} \times \pi \times (0.26 \times 10^{-3})^2 = 5.3 \times 10^{-8}$ m^2

To find the extension ΔL, rearrange the Young modulus equation

$E = \dfrac{TL}{A\Delta L}$ gives $\Delta L = \dfrac{TL}{AE} = \dfrac{58 \times 1.540}{5.3 \times 10^{-8} \times 2.1 \times 10^{11}} = 8.0 \times 10^{-3}$ m

Summary questions

The Young modulus of steel = 2.1×10^{11} Pa

1 A metal wire of diameter 0.26 mm and of unstretched length 1.524 m was suspended vertically from a fixed point. When a 48 N weight was suspended from the lower end of the wire, the wire stretched by an extension of 6.6 mm. Calculate the Young modulus of the wire material. *(3 marks)*

2 A vertical steel wire of diameter 0.22 mm and of length 1.500 m is fixed at its upper end, and has a weight of 32 N suspended from its lower end. Calculate:
   a the extension of the wire *(3 marks)*
   b the elastic energy stored in the wire. *(1 mark)*

3 A rectangular steel block of length 50 mm and cross-sectional area 3.4×10^{-4} m^2 is placed in a vice and its length is compressed by 0.13 mm when the vice is tightened. Calculate:
   a the compressive stress exerted on the block *(2 marks)*
   b the elastic energy stored in it. *(3 marks)*

1 a Define the density of a substance. *(1 mark)*

  b i A metal sphere has a diameter of 14.20 mm and a mass of 11.80 g. Calculate the density of the metal. *(2 marks)*

    ii The diameter value has an uncertainty of ±0.02 mm. The mass value has an uncertainty of ±0.04 g. Calculate the uncertainty in the density. *(3 marks)*

2 a Describe an experiment you would carry out to measure the density of the metal in a wire of diameter about 2 mm. *(3 marks)*

  b A square metal plate 50 mm by 50 mm in length and width has a mass of 14.20 g.

    i The following measurements of the plate thickness t were made at different places on the plate.

      2.10 mm   2.06 mm   2.14 mm   2.16 mm   2.08 mm

      Calculate the mean value of t and estimate the uncertainty in t. *(1 mark)*

    ii Calculate the volume V of the plate.

    iii Calculate the density ρ of the plate. *(3 marks)*

  c The uncertainty in the length of each side of the plate was +0.5 mm. The uncertainty in the mass of the plate was +0.10 g. Estimate the percentage uncertainty in ρ. *(3 marks)*

3 Figure 1 shows a steel spring suspended from a stand supporting a mass hanger of weight 0.25 N at its lower end. Different weights were added to the mass hanger and the extension of the spring was measured each time.

The results were plotted on the graph shown in Figure 2.

  a i Use the graph to determine the spring constant of the spring. *(2 marks)*

    ii Determine the length of the spring when it was unloaded and the mass hanger was removed. *(2 marks)*

  b Calculate the energy stored in the spring when its length was 450 mm. *(3 marks)*

4 a An elastic cord of unstretched length 320 mm has a cross-sectional area of 1.23 mm². The cord is stretched to a tension of 250 N. Assume that Hooke's law is obeyed for this range and that the cross-sectional area remains constant.

The Young modulus for the material of the cord = 3.9 GPa.

    i Calculate the extension of the cord at this tension. *(3 marks)*

    ii Calculate the work done to stretch the cord to this extension. *(1 mark)*

  b The cord is tested to investigate how much energy it absorbs when it is used to stop an object. Figure 3 shows the cord being struck at its midpoint with a hammer of mass 0.170 kg moving at an initial speed of 1.40 m s⁻¹. The hammer rebounds from the cord after hitting it.

    i Calculate the kinetic energy of the hammer before it hits the cord. *(1 mark)*

    ii The hammer rebounds from the cord at a speed of 1.15 m s⁻¹. Calculate the percentage of the kinetic energy of the hammer that it regains after the rebound. *(2 marks)*

    iii Explain why the hammer does not regain all its kinetic energy when it rebounds. *(2 marks)*

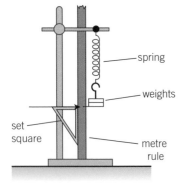

▲ **Figure 1**

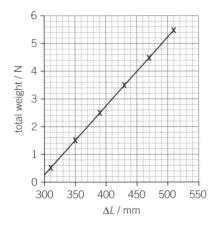

▲ **Figure 2**

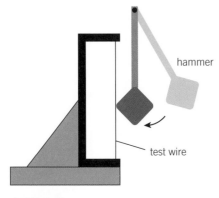

▲ **Figure 3**

5 A material in the form of a wire, 2.92 m long and with a diameter of 0.71 mm, is suspended from a support so that it hangs vertically. Different weights are suspended from its lower end and the extension of the wire is measured for each weight. The table shows the extension of the wire for each load as the weight is increased.

Load / N	0	9.8	19.6	29.4	39.2	49.0	58.8
Extension / mm	0	0.38	0.74	1.15	1.50	1.91	2.27

a i Plot a graph of load (on the *y*-axis) against extension (on the *x*-axis). *(3 marks)*

ii Use the graph to determine a value of the Young modulus for the material of the wire. *(5 marks)*

b The material of the wire has an ultimate tensile stress of 530 MPa. Calculate the maximum load the wire could support without breaking. *(1 mark)*

6 Figure 4 shows stress–strain curves for three different materials P, Q, and R up to the fracture point of each material. Answer the following questions about Figure 4, giving a reason for your answer to each question.

a Which of the three materials has the largest Young modulus? *(2 marks)*

b Which of the three materials is the most: **i** brittle **ii** ductile? *(4 marks)*

c Which material stores the most energy per unit volume up to its limit of proportionality? *(2 marks)*

7 A lift of weight 10 000 N is designed to carry a maximum passenger load of 6300 N. The lift is supported by six parallel steel cables, each of diameter 6.0 mm and of length 35.0 m.

Young modulus of steel = 210 GPa.

a Calculate the extension of each cable when the lift is moving upwards at a steady velocity and it is fully loaded. *(5 marks)*

b When the lift starts to ascend, it accelerates to a speed of 0.72 m s^{-1} in 8.0 s.

i Calculate the average acceleration of the lift during this time. *(1 mark)*

ii Estimate the additional extension of each cable during this time. *(4 marks)*

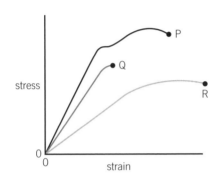

▲ **Figure 4**

12.1 Current and charge
12.2 Potential difference and power

Specification reference: 3.5.1.1

Electrical conduction

An electric current is the rate of flow of charge in the wire or component. The current is due to the passage of **charge carriers**.

- In a metallic conductor, the charge carriers are conduction electrons (i.e., electrons in the metal not attached to an atom). When a voltage is applied across the metal, these conduction electrons are attracted towards the positive terminal of the metal.

- In an insulator, each electron is attached to an atom and cannot move away from the atom. When a voltage is applied across an insulator, no current passes through the insulator, because no electrons can move through the insulator.

- In a semiconductor, the number of charge carriers increases with an increase of temperature. The resistance of a semiconductor therefore decreases as its temperature is raised.

- In a solution of salt, the charge carriers are positive and negative ions which are charged atoms or charged groups of atoms (i.e., molecules).

Charge flow

For a current I, the charge flow ΔQ in time Δt is given by $\Delta Q = I\,\Delta t$

For charge flow ΔQ in a time interval Δt, the current I is given by $I = \dfrac{\Delta Q}{\Delta t}$

The unit of current is the *ampere* (A), which is defined in terms of the magnetic force between two parallel wires when they carry the same current. The symbol for current is I.

The unit of charge is the *coulomb* (C), equal to the charge flow in one second when the current is one ampere. The symbol for charge is Q.

The magnitude of the charge of the electron $e = 1.60 \times 10^{-19}$ C. The convention for the direction of current in a circuit is from positive (+) to negative (−).

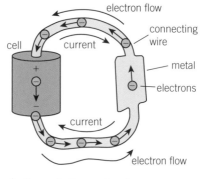

▲ **Figure 1** *Convention for current*

Energy and potential difference

Potential difference is defined as the work done (or energy transfer) per unit charge.

The unit of pd is the volt, which is equal to 1 joule per coulomb.

If work W is done when charge Q flows through the component, the pd across the component, V, is given by $V = \dfrac{W}{Q}$

The electromotive force (emf) of a source of electricity is defined as the electrical energy produced per unit charge passing through the source. The unit of emf is the volt, the same as the unit of pd.

For a source of emf ε in a circuit, the electrical energy produced when charge Q passes through the source = $Q\varepsilon$. The charge carriers are forced round the circuit and they transfer energy to the components in the circuit.

Revision tip
Remember 1 volt = 1 joule per coulomb.

Electrical power and current

Consider a component or device that has a potential difference V across its terminals and a current I passing through it. In time Δt:

- the work done W by the charge carriers = $IV\,\Delta t$
- the electrical power P supplied to the device = IV
- the unit of power is the watt (W). Therefore, one volt is equal to one watt per ampere.

Synoptic link

You will meet emf in more detail in Topic 13.3, Electromotive force and internal resistance.

Summary questions

$e = 1.60 \times 10^{-19}\,C$

1 In an electron beam experiment, the beam current was 1.2 mA for 300 s. Calculate:
   a the charge flowing along the beam   (*1 mark*)
   b the number of electrons that passed along the beam.   (*1 mark*)

2 A 230 V microwave oven has a power rating of 800 W. Calculate:
   a the current taken by the appliance   (*1 mark*)
   b the energy transfer to the appliance in 1 minute.   (*1 mark*)

3 A battery has an emf of 12 V and negligible internal resistance. It is capable of delivering a charge of 3.6×10^6 C. Calculate:
   a how long the battery will last after it has been fully charged when the current from it is 13 A   (*1 mark*)
   b the maximum energy it could deliver without being recharged.   (*1 mark*)

12.3 Resistance

Specification reference: 3.5.1.1; 3.5.1.3

Reminder about prefixes

▼ **Table 1** *Prefixes*

Prefix	Symbol	Value
Nano	n	10^{-9}
Micro	μ	10^{-6}
Milli	m	10^{-3}
Kilo	k	10^{3}
Mega	M	10^{6}
Giga	G	10^{9}

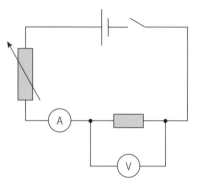

▲ **Figure 1** *Measuring resistance*

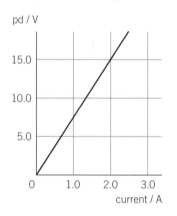

▲ **Figure 2** *Graph of pd versus current for a resistor*

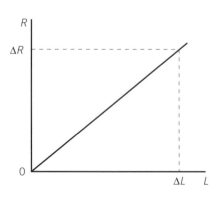

▲ **Figure 3** *Graph of resistance against length for a wire*

Definitions and laws

The resistance of a component in a circuit is caused by the repeated collisions between the charge carriers passing through the component and the positive ions in the component vibrating about fixed positions in the component.

The **resistance** of any component is defined as

$$\text{\textbf{the pd across the component}} \over \text{\textbf{the current through it}}$$

For a component which passes current I when the pd across it is V, its resistance R is given by the equation

$$R = \frac{V}{I}$$

The unit of resistance is the ohm (Ω), which is equal to 1 volt per ampere.

Rearranging the above equation gives $V = IR$ or $I = \frac{V}{I}$.

Measurement of resistance

The resistance of a resistor can be measured using the circuit shown in Figure 1.

To investigate the variation of current with pd, the variable resistor is adjusted in steps. At each step, the current and pd are recorded from the ammeter and voltmeter, respectively. The measurements can then be plotted on a graph of pd against current, as shown in Figure 2 (or for current against pd – see Topic 12.4, Components and their characteristics).

Ohm's law states that the pd across a metallic conductor is proportional to the current through it, provided the physical conditions do not change.

The graph for a resistor is a straight line through the origin so the pd across the resistor is proportional to the current. In this case, the resistance is equal to the gradient of the graph.

Resistivity

The resistivity ρ of a sample of material of length L, uniform cross-sectional area A, and resistance R is given by

$$\textbf{resistivity } \rho = \frac{RA}{L}$$

Note:

1   The unit of resistivity is the ohm metre (Ω m).

2   Rearranging the above equation gives $R = \rho \frac{L}{A}$.

To determine the resistivity of a wire:

- Measure the diameter of the wire d using a micrometer at several different points along the wire to give a mean value for d then calculate its cross-sectional area ($A = \frac{\pi d^2}{4}$).

- Measure the resistance R of different lengths L of wire to plot a graph of R against L, see Figure 3. The resistivity of the wire is given by the graph gradient × A.

Superconductivity

A **superconductor** is a wire or a device made of material that has zero resistivity at and below a **critical temperature** that depends on the material. This property of the material is called **superconductivity**. This critical temperature is also called its **transition temperature**. A superconductor material loses its superconductivity if its temperature is raised above its critical temperature.

Superconductors are used to make high-power electromagnets that generate very strong magnetic fields in devices such as magnetic resonance scanners and particle accelerators. Superconductors with higher transition temperatures could be used in new applications such as lightweight electric motors and power cables that transfer electrical energy without energy dissipation.

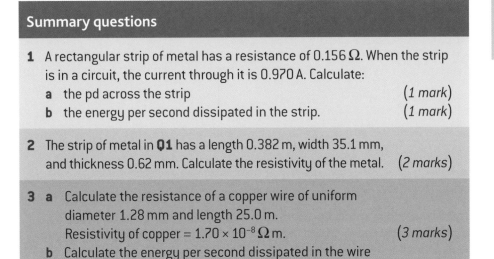

▲ **Figure 4** *Resistivity of a superconductor versus temperature near the transition temperature*

Revision tip
Resistivity is a property of a material. Don't confuse resistivity and resistance.

Summary questions

1 A rectangular strip of metal has a resistance of $0.156\,\Omega$. When the strip is in a circuit, the current through it is $0.970\,A$. Calculate:
   a the pd across the strip *(1 mark)*
   b the energy per second dissipated in the strip. *(1 mark)*

2 The strip of metal in **Q1** has a length $0.382\,m$, width $35.1\,mm$, and thickness $0.62\,mm$. Calculate the resistivity of the metal. *(2 marks)*

3 a Calculate the resistance of a copper wire of uniform diameter $1.28\,mm$ and length $25.0\,m$.
      Resistivity of copper $= 1.70 \times 10^{-8}\,\Omega\,m$. *(3 marks)*
   b Calculate the energy per second dissipated in the wire when the current through it is $13.0\,A$. *(1 mark)*

Investigating the characteristics of different components

To measure the variation of current with pd for a component, use (as shown in Figure 1) either:

- a potential divider to vary the pd from zero, or
- a variable resistor to vary the current to a minimum.

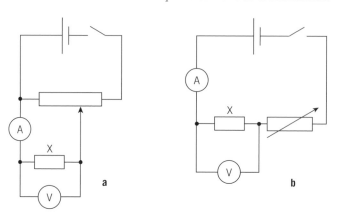

▲ **Figure 1** *Investigating component characteristics* **a** *Using a potential divider,* **b** *Using a variable resistor*

The advantage of using a potential divider is that the current through the component and the pd across it can be reduced to zero. This is not possible with a variable resistor circuit.

The measurements for each type of component are usually plotted as a graph of current (on the *y*-axis) against pd (on the *x*-axis). Typical graphs for a wire, a filament lamp, and a thermistor are shown in Figure 2. Note that the measurements are the same, regardless of which way the current passes through each of these components.

In both circuits, an ammeter sensor and a voltmeter sensor connected to a data logger could be used to capture data (i.e., to measure and record the readings) which could then be displayed directly on an oscilloscope or a computer.

- A wire gives a straight line through the origin. This means that the resistance of the wire does not change when the current changes. In this case, the gradient of the line is equal to $\frac{1}{\text{resistance } R \text{ of the wire}}$.
- A filament bulb gives a curve with a decreasing gradient because its resistance increases as it becomes hotter.
- A thermistor at constant temperature gives a straight line. The higher the temperature, the greater the gradient of the line, as the resistance falls with increase of temperature. The same result is obtained for a **light-dependent resistor** in respect of light intensity.

The diode

To investigate the characteristics of the diode, one set of measurements is made with the diode in its forward direction (i.e., forward biased) and another set with it in its reverse direction (i.e., reverse biased). The current is very small when the diode is reverse biased and can only be measured using a milliammeter.

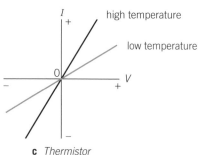

a *Wire*

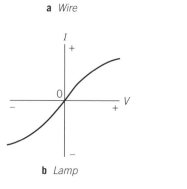

b *Lamp*

c *Thermistor*

▲ **Figure 2** *Current versus pd for different components*

Typical results for a silicon diode are shown in Figure 3. A silicon diode conducts easily in its forward direction above a pd of about 0.6 V and hardly at all below 0.6 V or in the opposite direction. This voltage is called the 'forward voltage' of the diode.

Resistance and temperature

1   The resistance of a metal increases with increase of temperature. This is because the positive ions in the conductor vibrate more when its temperature is increased so the charge carriers (conduction electrons) cannot pass through the metal as easily. A metal has a **positive temperature coefficient** because its resistance increases with increase of temperature.

2   The resistance of an intrinsic semiconductor decreases with increase of temperature. This is because the number of charge carriers (conduction electrons) increases when the temperature is increased. A thermistor made from an intrinsic semiconductor therefore has a **negative temperature coefficient**.

The resistance of the thermistor decreases non-linearly with increase of temperature, whereas the resistance of the metal wire increases much less over the same temperature range.

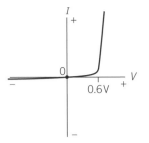

▲ **Figure 3** *Current versus pd for a diode*

Revision tip

When current passes through a forward-biased silicon diode, the pd across the diode is about 0.6 V.

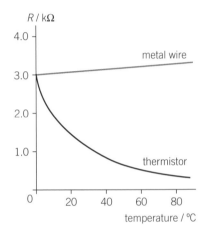

▲ **Figure 4** *Resistance variation with temperature for a thermistor and a metal wire*

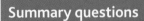

Summary questions

1   **a**   Sketch a graph of current I (on the y-axis) against pd V for a filament bulb for positive and negative values of V.   *(1 mark)*

   b   Describe how and explain why the resistance of the filament bulb changes as the current is increased from zero.   *(4 marks)*

2   A light-dependant resistor has a resistance of $100\,000\,\Omega$ in darkness and $400\,\Omega$ in daylight. It is connected in series with an ammeter and a 4.5 V cell. Calculate the ammeter reading when the thermistor is:

   a   **i** in darkness   **ii** in daylight.   *(2 marks)*

   b   Explain why the resistance in darkness is much greater than in daylight.   *(2 marks)*

3   A diode is connected in series with a cell, an ammeter, and a resistor.

   a   Draw a circuit diagram for this circuit, showing the diode in its forward direction.   *(1 mark)*

   b   State and explain how the ammeter reading would change if the resistor was replaced by a resistor with greater resistance.   *(2 marks)*

Chapter 12 Practice questions

1 A 6.0 V battery is connected across a wire-wound resistor. The current in the resistor is 2.6 A.

  a The wire has a diameter of 0.38 mm and a resistivity of $4.8 \times 10^{-7}\,\Omega\,\text{m}$. Calculate the length of the wire. *(4 marks)*

  b i Calculate the total charge that flows past a point in the conductor in 10 minutes.

   ii Calculate the energy transferred to the resistor by the electric current in 10 minutes. *(2 marks)*

  c Describe the energy transfers to and from the resistor due to the current. *(3 marks)*

2 Figure 1 shows a graph of V against I for a 12 V filament lamp.

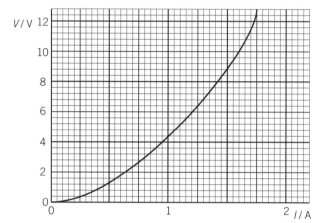

▲ **Figure 1**

  a Calculate the resistance of the lamp at: **i** 6.0 V **ii** 12.0 V. *(2 marks)*

  b When the pd across the lamp is 12.0 V, calculate:

   i the energy supplied to the lamp in 300 s

   ii the charge that flows through the lamp in 300 s. *(2 marks)*

  c Show that the power dissipated by the lamp at 12.0 V is about 3 times the power dissipated at 6.0 V. *(2 marks)*

3 The following measurements were made in an investigation to measure the resistivity of the material of a certain wire.

Pd across the wire / V	0.00	1.10	2.05	3.90	4.10	4.95
Length of the wire / mm	198	400	603	796	998	1202

Current = 0.41 A

Diameter of wire = 0.26 mm

  a Plot a graph of the pd against the length of the wire. *(3 marks)*

  b i Show that the pd, V, across the wire varies with the length, L, according to the equation

$$V = \frac{\rho I L}{A}$$

   where ρ is the resistivity of the wire, A is its area of cross section, and I is the current in the wire

   ii Use the graph to calculate the resistivity of the material of the wire. *(6 marks)*

  c Discuss how the graph would have differed if a thinner wire of the same material and the same current had been used. *(2 marks)*

4 a Describe, with the aid of a circuit diagram, how you would obtain an estimate of the ratio of the resistance of a light-dependent resistor (LDR) in darkness to when it is daylight. The apparatus available includes a battery, a variable resistor, an ammeter, and a voltmeter. *(5 marks)*

b A student uses a light-dependent resistor in series with a 4.5 V battery and an ammeter as a light meter as shown in Figure 2

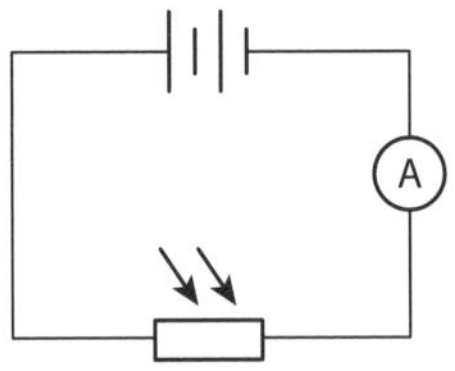

  i Explain how the ammeter reading changes when the light intensity incident on the LDR increases. *(2 marks)*

  ii When the LDR is in daylight, the ammeter reads 1.2 mA. Calculate the resistance of the LDR in this situation. *(1 mark)*

  iii Explain why the resistance of an LDR is much greater in darkness than in daylight. *(3 marks)*

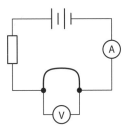

▲ **Figure 2**

5 a Explain what is meant by a forward voltage of a diode. *(1 mark)*

b The current in a light-emitting diode (LED) in a circuit is 0.17 A when forward voltage across it is 2.2 V.

  i Calculate the charge that flows through it in 10 ms.

  ii Calculate the power supplied to the LED described above. *(2 marks)*

c The LED in part **b** emits photons of energy 3.5×10^{-19} J.

  i Assuming all the energy supplied to the LED is emitted as light, estimate the maximum number of photons per second that the LED could emit when the current through it is 0.17 A.

  ii Give one reason why the LED emits fewer photons per second than calculated in part **c i**. *(2 marks)*

6 a i What is meant by the transition temperature of a superconductor? *(1 mark)*

  ii A certain superconducting material M has a transition temperature of 90 K. Sketch a graph to show how the resistivity of M changes when its temperature is increased from below to above 90 K. *(2 marks)*

b Figure 3 shows a sample of M in the form of a wire in series with a resistor in a circuit.

  i When the temperature of M is below its transition temperature, explain why the potential difference across M is zero even though the current through it is non-zero. *(1 mark)*

  ii Explain how the ammeter and voltmeter readings would change if the temperature of M is raised above its transition temperature. *(2 marks)*

▲ **Figure 3**

7 The heating element of an electric heater consists of a metal wire W coiled round a ceramic tube as shown in Figure 4.

tube   wire

▲ **Figure 4**

a The wire has a diameter of 0.30 mm and the resistivity of the wire metal is 5.6×10^{-8} Ω m. When the pd across the element is 230 V, the current in it is 10.8 A. Calculate the length of W. *(4 marks)*

b A wire X of the same material has a diameter of 0.60 mm.

  i Calculate the length of X that would have the same resistance as W. *(3 marks)*

  ii Discuss whether X would be more suitable than W as the heating element in the electric heater. *(4 marks)*

13.1 Circuit rules

Specification reference: 3.5.1.4

Revision tip

1 volt = 1 joule per coulomb

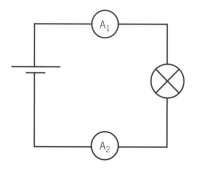

▲ **Figure 1** *Components in series*

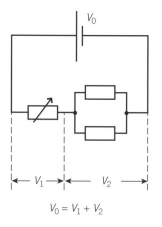

$$V_0 = V_1 + V_2$$

▲ **Figure 2** *Two components in parallel*

Synoptic link

Sources of emf usually possess some internal resistance. See Topic 13.3, Electromotive force and internal resistance.

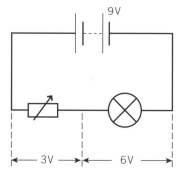

▲ **Figure 3** *The loop rule*

Current rules

1 At any junction in a circuit, the total current leaving the junction is equal to the total current entering the junction.

2 The current entering a component is the same as the current leaving it.

3 For two or more components in series, the current in them is the same.

These rules hold because the rate of charge flowing into a junction or a component is always equal to the rate flowing out.

Potential difference rules

The potential difference (abbreviated as pd), or voltage, between any two points in a circuit is defined as the energy transfer per coulomb of charge that flows from one point to the other.

1 For two or more components in series, the total pd across all the components is equal to the sum of the potential differences across each component.

2 The pd across components in parallel is the same.

3 For any complete loop of a circuit, the sum of the emfs round the loop is equal to the sum of the potential drops around the loop.

The above statements follow from the conservation of energy. For example, in Figure 3, if the pd across the light bulb is 6 V, the pd across the variable resistor is 3 V (= 9 V – 6 V). So every coulomb of charge leaves the battery with 9 J of energy and supplies 3 J to the variable resistor and 6 J to the bulb.

Summary questions

1 A battery which has an emf of 9.0 V and negligible internal resistance is connected in series with a variable resistor and a 6.0 V 24 W light bulb.
   a Sketch the circuit diagram. *(1 mark)*
   b The variable resistor is adjusted so the light bulb is at normal brightness. Calculate: **i** the current through the light bulb **ii** the pd across the variable resistor **iii** the power supplied by the battery. *(3 marks)*

2 A 4.5 V battery of negligible internal resistance is connected in parallel with a 4.5 V 0.5 W torch bulb and a 4.5 V 3.0 W torch bulb.
   a Sketch the circuit diagram for this circuit. *(1 mark)*
   b Calculate the current through each torch bulb. *(2 marks)*
   c Show that the energy supplied by the battery each second equals the energy supplied to the torch bulbs each second. *(3 marks)*

3 A 9.0 V battery of negligible internal resistance is connected in series with an ammeter, a 1.5 kΩ resistor, and an unknown resistor R.
   a Sketch the circuit diagram. *(1 mark)*
   b The ammeter reads 2.0 mA. Calculate:
      i the pd across the 1.5 kΩ resistor
      ii the pd across R **iii** the resistance of R. *(3 marks)*

13.2 More about resistance

Specification reference: 3.5.1.4

Resistors in series

Resistors in series pass the same current. The total pd is equal to the sum of the individual pds. For two or more resistors R_1, R_2, R_3, and so on, in series, the theory can easily be extended to show that the total resistance is equal to the sum of the individual resistances.

$$R = R_1 + R_2 + R_3 + \cdots$$

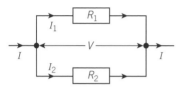

▲ **Figure 1** Resistors in series

Resistors in parallel

Resistors in parallel have the same pd. The current through a parallel combination of resistors is equal to the sum of the individual currents.

For two or more resistors R_1, R_2, R_3, and so on, in parallel, the theory can easily be extended to show that the total resistance R is given by

$$\frac{1}{R} = \frac{1}{R_1} + \frac{1}{R_2} + \frac{1}{R_3} + \cdots$$

▲ **Figure 2** Resistors in parallel

Resistance heating

The heating effect of an electric current in any component is due to the resistance of the component. The charge carriers repeatedly collide with the positive ions of the conducting material. There is a net transfer of energy from the charge carriers to the positive ions as a result of these collisions.

When a current I passes through a component of resistance R, the pd across it is V.

$$\text{the rate of heat transfer} = I^2 R = IV = \frac{V^2}{R}$$

Summary questions

1   A $4.0\,\Omega$ resistor and a $12.0\,\Omega$ resistor are connected in parallel with each other. The parallel combination is connected in series with a $6.0\,V$ battery and a $2.0\,\Omega$ resistor. Assume the battery itself has negligible internal resistance.
    a   Sketch the circuit diagram.   (1 mark)
    b   Calculate: **i** the total resistance of the circuit
        ii the battery current.   (3 marks)
    c   Calculate the power supplied to the $4.0\,\Omega$ resistor.   (2 marks)

2   A $2.0\,\Omega$ resistor and a $10.0\,\Omega$ resistor are connected in series with each other. The series combination is connected in parallel with a $4.0\,\Omega$ resistor and in parallel with a $9.0\,V$ battery of negligible internal resistance.
    a   Sketch the circuit diagram.   (2 marks)
    b   Calculate: **i** the total resistance of the circuit
        ii the battery current.   (3 marks)
    c   Calculate the power supplied to each resistor.   (3 marks)

3   Calculate the resistance of a heating element designed to operate at 3 kW and 230 V.   (2 marks)

Worked example

The pd across a $1000\,\Omega$ resistor in a circuit was measured at $6.0\,V$. Calculate the electrical power supplied to the resistor.

Solution

Current $I = \dfrac{V}{R} = \dfrac{6.0\,V}{1000\,\Omega} = 6.0\,mA$

Power $P = I^2 R = (6.0 \times 10^{-3})^2 \times 1000$
$= 0.036\,W$

13.3 Electromotive force and internal resistance

Specification reference: 3.5.1.6

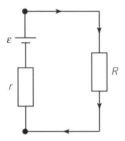

▲ **Figure 1** *Emf and internal resistance*

> **Revision tip**
>
> Multiplying each term of the equation opposite by the cell current I gives
>
> **power supplied by the cell**
>
> $$I\varepsilon = I^2R + I^2r$$
>
> In other words, the power supplied by the cell = the power delivered to R + the power wasted in the cell due to its internal resistance.

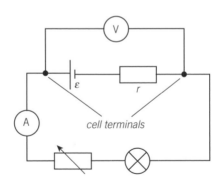

▲ **Figure 2** *Measuring internal resistance. Note that the lamp (or a fixed resistor) limits the maximum current that can pass through the cell*

Internal resistance

- The **electromotive force** (emf, symbol ε) of the source is the electrical energy per unit charge produced by the source. If electrical energy E is given to a charge Q in the source,

$$\varepsilon = \frac{E}{Q}$$

- The *pd across the terminals* of the source is the electrical energy per unit charge delivered by the source when it is in a circuit. The terminal pd is less than the emf whenever current passes through the source. The difference is due to the internal resistance of the source.

The internal resistance of a source is the loss of potential difference per unit current in the source when current passes through the source.

When a cell of emf ε and internal resistance r is connected to an external resistor of resistance R, as shown in Figure 1, the current through the cell, $I = \frac{\varepsilon}{R + r}$. Therefore,

$$\varepsilon = IR + Ir$$

Measurement of internal resistance

The circuit shown in Figure 2 is used to measure the terminal pd for different values of current. The current is changed by adjusting the variable resistor. The measurements of terminal pd and current for a given cell may be plotted on a graph, as shown in Figure 3.

The graph is a straight line with a negative gradient. This can be seen by rearranging the equation $\varepsilon = IR + Ir$ to become $IR = \varepsilon - Ir$. Because the terminal pd $V = IR$, then

$$V = \varepsilon - Ir$$

Comparing this with the equation for a straight line, $y = mx + c$, a graph of V on the y-axis against I on the x-axis gives a straight line with

- gradient $= -r$
- y-intercept $= \varepsilon$.

Summary questions

1 A battery of emf 9.0 V and internal resistance $0.5\,\Omega$ is connected to a $4.5\,\Omega$ resistor. Calculate:
   a the total resistance of the circuit (*1 mark*)
   b the current through the battery (*1 mark*)
   c the pd across the cell terminals. (*1 mark*)

2 A cell of emf 2.0 V and internal resistance $0.5\,\Omega$ is connected to a $3.5\,\Omega$ resistor. Calculate:
   a the current (*2 marks*)
   b the power delivered to the $3.5\,\Omega$ resistor (*1 mark*)
   c the power wasted in the cell. (*1 mark*)

3 The pd across the terminals of a cell was 1.05 V when the current from the cell was 0.25 A, and 1.35 V when the current was 0.10 A. Calculate:
   a the internal resistance of the cell (*3 marks*)
   b the cell's emf. (*1 mark*)

13.4 More circuit calculations

Specification reference: 3.5.1.4; 3.5.1.6

Circuits with a single cell and one or more resistors

1 The **cell current** = $\dfrac{\text{cell emf}}{\text{total circuit resistance}}$

The total circuit resistance includes internal resistance if that is not negligible.

2 The **pd across each resistor in series with the cell = current × the resistance of each resistor.**

Circuits with two or more cells in series

The current through the cells is calculated by dividing the overall (net) emf by the total resistance.

• The overall emf is the difference between the sum of the emfs in each direction.

• The total internal resistance is the sum of the individual internal resistances.

Circuits with identical cells in parallel

For a circuit with n identical cells in parallel:

• the cells act as a source of emf ε and internal resistance $\dfrac{r}{n}$

• the current through each cell = $\dfrac{I}{n}$, where I is the total current supplied by the cells.

Diodes in circuits

Assume that a silicon diode has:

• a forward pd of 0.6 V whenever a current passes through it

• infinite resistance in the reverse direction or at pds less than 0.6 V in the forward direction.

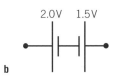

2.0V   1.5V

a

2.0V   1.5V

b

▲ **Figure 1** *Cells in series*

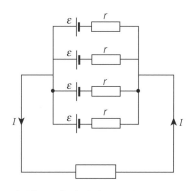

▲ **Figure 2** *Cells in parallel*

> **Revision tip**
>
> In problems on dc circuits that include filament lamps, if you are given data for the pd V and power P of the lamp, its current $I = \dfrac{P}{V}$.

> **Worked example**
>
> A silicon diode is connected in its forward direction in series with a 5.0 V cell of negligible internal resistance and a 15 kΩ resistor, as in Figure 3. Calculate the current through the diode.
>
> ### Solution
>
> The pd across the diode is 0.6 V because it is forward-biased. Therefore, the pd across the resistor is 4.4 V (= 5.0 V − 0.6 V). The current through the resistor is therefore 2.9×10^{-4} A (= 4.4 V ÷ 15 000 Ω).

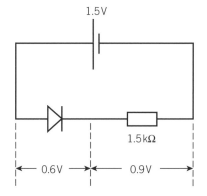

1.5V

1.5 kΩ

0.6 V   0.9 V

▲ **Figure 3** *Using a diode*

> **Summary questions**
>
> 1 A battery of emf 12.0 V with an internal resistance of 1.55 Ω is connected in series with a variable resistor and two 6.0 V 6.0 W lamps X and Y in parallel with each other. The variable resistor is adjusted so the lamps are at normal brightness.
>   **a** Sketch the circuit diagram.   (*1 mark*)
>   **b** Calculate:
>   **i** the current through each lamp   (*1 mark*)
>   **ii** the battery current   (*1 mark*)
>   **iii** the pd across the variable resistor.   (*2 marks*)

> **Synoptic link**
>
> A solar panel consists of parallel rows of identical solar cells in series. For more about solar panels, see 'Renewable energy' in Chapter 10, Work, energy, and power.

13.5 The potential divider

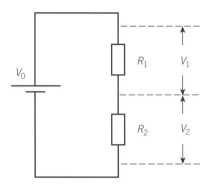

▲ **Figure 1** *A potential divider*

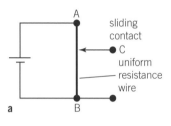

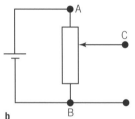

▲ **Figure 2** *Potential dividers used to supply a variable pd* **a** *A linear track using resistance wire* **b** *Circuit symbol*

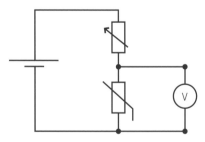

▲ **Figure 3** *A temperature sensor*

The theory of the potential divider

A **potential divider** consists of two or more resistors in series with each other and with a source of fixed potential difference.

To supply a fixed pd

Two resistors R_1 and R_2 in series are connected to a source of fixed pd V_0, as shown in Figure 1.

The ratio of the pds across each resistor is equal to the resistance ratio of the two resistors.

$$\frac{V_1}{V_2} = \frac{R_1}{R_2}$$

To supply a variable pd

The source pd is connected to a fixed length of uniform resistance wire. A sliding contact on the wire moved along the wire gives a variable pd between the contact and one end of the wire.

Sensor circuits

A *sensor circuit* produces an output pd which changes as a result of a change of a physical variable such as temperature or light intensity.

A *temperature sensor* consists of a potential divider made using a thermistor and a variable resistor, as in Figure 3. If the temperature of the thermistor is then raised, its resistance falls, so the pd across it falls.

Summary questions

1. **a** Sketch a circuit diagram to vary the brightness of a 12 V light bulb between zero and a maximum using a variable potential divider and a 12 V battery. *(1 mark)*
   b Explain why the same range of brightness could not be obtained if the light bulb was connected in series with the battery and a variable resistor. *(3 marks)*

2. **a** A potential divider consists of an 8.0 Ω resistor, R, in series with a 12.0 Ω resistor and a 6.0 V battery of negligible internal resistance. Calculate:
   i the current *(2 marks)*
   ii the pd across each resistor. *(3 marks)*
   b The 12.0 Ω resistor is replaced by a thermistor with a resistance of 16.0 Ω at 20 °C and 6.0 Ω at 100 °C. Calculate the pd across R at:
   i 20 °C **ii** 100 °C. *(4 marks)*

3. A light sensor consists of a 4.5 V battery of negligible internal resistance, an LDR, and a 2.2 kΩ resistor in series with each other. A voltmeter is connected in parallel with the resistor. When the LDR is in darkness, the voltmeter reads 1.2 V.
   a Sketch the circuit diagram for this arrangement. *(1 mark)*
   b Calculate: **i** the pd across the LDR **ii** the resistance of the LDR when the voltmeter reads 1.2 V. *(3 marks)*
   c Describe and explain how the voltmeter reading changes when the LDR is exposed to daylight. *(2 marks)*

Chapter 13 Practice questions

1 A circuit consists of two resistors of resistance 3.0 Ω and 6.0 Ω in parallel with each other and connected in series with an ammeter and a 6.0 V battery which has an internal resistance of 3.0 Ω, see Figure 1.

    a Calculate: **i** the current through the ammeter **ii** the pd across the 3.0 Ω resistor. *(4 marks)*

    b When one of the resistors is disconnected from the circuit, describe and explain how the pd across the battery changes. *(3 marks)*

2 A 9.0 V battery of negligible internal resistance is connected in series with a silicon diode in its forward direction and a 6.8 kΩ resistor, see Figure 2.

    a Calculate: **i** the pd across the 6.8 kΩ resistor **ii** the battery current. *(3 marks)*

    b Another 6.8 kΩ resistor is connected in parallel with the diode to 6.8 kΩ resistor. Explain the effect this has on the battery current. *(2 marks)*

3 a With the aid of a circuit diagram, describe an experiment to measure the internal resistance of a battery. *(5 marks)*

    b A 6.0 V battery of internal resistance r is connected across a 3.0 Ω resistor in parallel with a 9.0 Ω resistor, see Figure 3. The potential difference across the battery is 5.4 V.

       i Calculate the current in the battery.

       ii Calculate the internal resistance of the battery.

       iii Calculate the ratio $\frac{\text{power dissipated in the internal resistance}}{\text{power supplied by the battery}}$. *(5 marks)*

4 Figure 4 shows a circuit diagram for a dc power supply connected to a mobile phone battery B in order to charge the battery. The terminals of the same polarity are connected together to achieve this.

    a The power supply has an emf of 5.0 V and internal resistance 0.9 Ω. When charging begins, the battery has an emf of 2.0 V and internal resistance r.

       i Calculate the net emf of the circuit when charging begins.

       ii The initial charging current is 0.50 A. Calculate the value of r. *(3 marks)*

    b **i** As the battery charges, its emf increases and the current decreases. Explain why the current decreases.

       ii The battery takes 3.0 hours to charge and can store 1.6 kC of charge. Estimate the average current in the battery when it is recharged. *(3 marks)*

5 A solar panel connected to an appliance produces a potential difference of 6.0 V and a current of 3.0 A when it is illuminated by light of constant intensity.

    a Calculate the energy supplied by the solar panel in 1 hour. *(1 mark)*

    b Each solar cell in the panel produces a potential difference of 1.2 V and a current of 1.0 A.

       i Calculate how many cells are in the panel and sketch the circuit diagram to show how they are connected together. *(3 marks)*

       ii When the appliance is disconnected from the panel without changing the incident light intensity, the potential difference across its terminals increases to 6.3 V. Calculate the effective internal resistance of the panel at this light intensity. *(3 marks)*

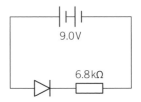

▲ Figure 1

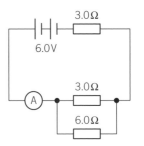

▲ Figure 2

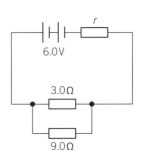

▲ Figure 3

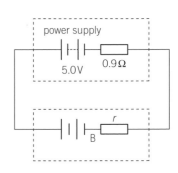

▲ Figure 4

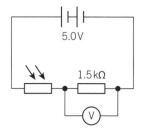

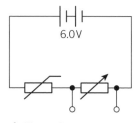

▲ Figure 5

6 Figure 5 shows a potential divider circuit that includes a light-dependent resistor (LDR) in series with a 1.5 kΩ resistor R. The battery has an emf of 5.0 V and has negligible internal resistance.

a Explain what is meant by a potential divider. (*2 marks*)

b The potential difference V_0 across the resistor R is 4.2 V when the LDR is exposed to light of a certain intensity. Calculate the LDR resistance at this light intensity. (*3 marks*)

c State and explain what happens to the output voltage when the light intensity incident on the LDR is reduced. (*2 marks*)

d Describe how you would use the above circuit with a voltmeter connected across the output terminals to investigate how the light intensity from a lamp varies as the LDR is moved around the lamp. (*6 marks*)

7 The potential divider in Figure 6 is to be used to monitor the temperature of a cold storage room.

a State and explain how the potential difference across the variable resistor changes when the temperature of the thermistor is increased. (*2 marks*)

b i The thermistor T has a resistance of 2.7 kΩ at −5 °C. Calculate the resistance of the variable resistor that would give an output voltage of 2.0 V. (*3 marks*)

ii An alarm is to be connected to the output terminals of the potential divider. With the variable resistor set at the resistance calculated in part **b i**, the alarm switches on if the temperature of the thermistor increases above −5 °C. If the variable resistor is adjusted to a lower resistance, explain how this would affect the temperature at which the alarm switches on. (*4 marks*)

▲ Figure 6

Answers to practice questions

Chapter 1

1 a i 11 **ii** 12 **iii** 11 [1] *(1 mark)*

b Specific charge of the ion $= \dfrac{e}{23m_{nuc}}$ [1]

$= \dfrac{1.60 \times 10^{-19}\,\text{C}}{23 \times 1.67 \times 10^{-27}\,\text{kg}}$ [1]

$= 4.17 \times 10^{6}\,\text{C kg}^{-1}$ (where m_{nuc} is the mass of a nucleon) [1] *(3 marks)*

2 a Let N be the number of protons in the nucleus. Therefore, the specific charge $= \dfrac{Ne}{38m_{nuc}}$ [1]
$= 4.30 \times 10^{7}\,\text{C kg}^{-1}$.

Hence, $N = \dfrac{38 \times 1.67 \times 10^{-27} \times 4.3 \times 10^{7}}{1.60 \times 10^{-19}}$ [1]

$= 17$ [1] *(3 marks)*

b $^{38}_{17}\text{Cl} \rightarrow {}^{38}_{18}\text{Ar} + {}^{\ 0}_{-1}\beta$ [3] (1 mark for each correct term) *(3 marks)*

c Specific charge after emission $= \dfrac{18e}{38m_{nuc}}$

$= \dfrac{18 \times 1.60 \times 10^{-19}}{38 \times 1.67 \times 10^{-27}}$ [1] $= 4.54 \times 10^{7}\,\text{C kg}^{-1}$ [1] *(2 marks)*

3 a a = 206, b = 81, c = 4, d = 2 [1] for any 2 correct [1] for remaining 2 correct *(2 marks)*

b X = the strong nuclear force which acts between the nucleons in the nucleus [1]

c Y = the electromagnetic force (of repulsion) between the protons in the nucleus [1] *(2 marks)*

4 a $E = \dfrac{hc}{\lambda} = \dfrac{6.63 \times 10^{-34} \times 3.00 \times 10^{8}}{630 \times 10^{-9}} = 3.16 \times 10^{-19}\,\text{J}$ [1]

$= \dfrac{3.16 \times 10^{-19}}{1.60 \times 10^{-19}}\,\text{eV} = 1.97\,\text{eV}$ [1] *(2 marks)*

b Let N be the number of photons per second emitted. Therefore, $N E = 4.0 \times 10^{-3}\,\text{W}$

Hence, $N = \dfrac{4.0 \times 10^{-3}\,\text{W}}{3.16 \times 10^{-19}\,\text{J}}$ [1] $= 1.3 \times 10^{16}\,\text{s}^{-1}$ [1] *(2 marks)*

5 a It has an infinite range [1], it acts between charged particles and is attractive between oppositely charged particles and repulsive between particles with the same type of charge. [1] *(2 marks)*

b The force is due to the exchange of virtual photons between the two charged objects. [1] The photons are said to be virtual because they cannot be detected [1]. *(2 marks)*

6 a A positron is the antiparticle of the electron. [1] *(1 mark)*

b i A proton changes into a neutron and emits a positron in the process. So the number of protons decreases by one and the number of neutrons increases by 1. [1] *(1 mark)*

ii The weak nuclear force is responsible. The exchange particle is the W^{+} boson. [1] *(1 mark)*

iii a = proton, b = neutron [1], c = W^{+} boson [1], d and e = positron (β^{+} or e^{+}) and a neutrino (n) in either order. [1] *(3 marks)*

7 a A proton in the nucleus interacts with an inner shell electron by emitting a W^{+} boson which is absorbed by the inner shell electron [1]. As a result, the proton changes into a neutron and the electron changes into a neutrino. [1] *(2 marks)*

b See Figure 1: any 2 correct [1] remaining two correct [1] *(2 marks)*

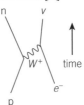

▲ **Figure 1**

8 a Pair production occurs when a photon of sufficient energy passing through matter near the nucleus of an atom [1] creates a particle and its corresponding antiparticle and ceases to exist. [1] An example is when a high-energy gamma photon creates an electron and a positron (or other correct example). [1] *(3 marks)*

b $2E_{ph} = 2 \times 0.51\,\text{MeV}$ [1]
Therefore, $E_{ph} = 0.51\,\text{MeV}$ [1] *(2 marks)*

c Although annihilation is a process that turns matter into radiation and pair production turns radiation into matter, they are not reverse processes [1] because pair production involves a single photon passing near a nucleus whereas annihilation occurs whenever and wherever a particle and its corresponding antiparticle meet. [1] In both processes, energy and momentum are conserved but in the case of pair production, the nucleus is necessary to enable momentum to be conserved whereas for annihilation two photons are always produced in order to conserve momentum. [1] *(3 marks)*

Chapter 2

1 a $\overline{\text{p}}$ **b** e^{+}, v_{e}, μ^{+} [1] **c** $\overline{\text{p}}$ **d** K^{-} [1] *(2 marks)*

2 a Hadrons can interact through the strong interaction. Leptons do not interact through the strong interaction. [1] *(1 mark)*

b i The weak interaction. [1]

ii The lepton numbers are −1 for the antimuon, the positron, and the muon antineutrino. The electron antineutrino has a lepton number of 1. The muon lepton number before the change is −1 and −1 after the change. Inserting these values into the table below shows that electron lepton number is conserved and muon lepton number is conserved.

	μ^{+}	$\rightarrow$	e^{+}	v_{e}	$\overline{v_{\mu}}$	
Electron lepton number	0	=	−1	+1	0	[1]
Muon lepton number	−1	=	0	0	−1	[1]

iii Difference: the muon decays, the electron does not. Similarity: both are negatively charged. *(3 marks)*

3 a udd [1], the charge of an up quark is $+\frac{2}{3}$ and the charge of a down quark is $-\frac{1}{3}$ so the total charge of the udd combination is 0. [1] *(2 marks)*

b There are 2 non-strange charged mesons. [1] Their quark-antiquark compositions in brackets are π^- ($\overline{u}d$) and π^+ ($u\overline{d}$). The antiparticle of each of these mesons is therefore the other meson in the pair. [1] *(2 marks)*

4 a i π^+ [1]

ii The charge of an up quark is $+\frac{2}{3}$ and the charge of a down antiquark is $+\frac{1}{3}$ so the total charge of the ($u\overline{d}$) combination is +1. [1] *(2 marks)*

b i X is composed of 3 quarks because it is a baryon. [1] Two of the quarks are strange quarks because the strangeness of X is −2 and a strange quark has a strangeness of −1. [1]

The charge of a strange quark is $-\frac{1}{3}$ so the two strange quarks provide a charge of $-\frac{2}{3}$. The remaining charge is $-\frac{1}{3}$ because X has a charge of −1. [1] So the remaining charge is provided by a down quark. Therefore, X's quark composition is dss. [1] *(4 marks)*

ii All baryons eventually decay into a proton so X will decay into a proton. *(1 mark)*

5 a A proton in the nucleus changes into a neutron by emitting a W^+ boson [1] which decays into a positron and an electron neutrino. [1] See Topic 2.4, Fig 2b. [1] *(3 marks)*

b Inserting values for baryon number, lepton number, and charge into the table below shows that electron lepton number is not conserved and muon lepton number is not conserved. So the decay will not happen. [1]

	μ+	→	e+	v_e	$\overline{v}_\mu$
Charge	+1	=	+1	0	0
Baryon number	0	=	0	0	0
Electron lepton number	0	≠	−1	−1	0
Muon lepton number	−1	≠	0	0	1

Q and B rows correct [1] lepton rows correct [1] *(3 marks)*

6 a Strangeness is conserved as the reaction is a strong interaction. The total initial strangeness is −1 so the total final strangeness must also be −1. Therefore, the strangeness of X is −1 as the π^- has zero strangeness. [1] The charge of X is +1 as charge is conserved, the total initial charge is zero, and the π^- has a charge of −1. [1] The baryon number of the π^- meson and of the K^- meson is zero. Because baryon number is conserved and

the baryon number of the proton is +1, the baryon number of X must also be +1. [1] *(3 marks)*

b As X includes a strange quark in its composition, then another particle containing a strange antiquark must be produced in order to conserve strangeness (as the collision is a strong interaction). [1] So the other particle could not be a π meson which is a non-strange particle. [1] *(2 marks)*

7 a Similarity: both positively charged. Difference: an antimuon is an antiparticle, proton is a particle (or an antimuon is a lepton and a proton is a hadron). [1]

b Similarity: both are uncharged (or both are mesons/hadrons). Difference: a K^0 meson can decay into π mesons whereas a π^0 meson does not decay into K mesons. [1]

c Similarity: both are hadrons. Difference: the π^+ meson is charged whereas the neutron is uncharged (or the π^+ meson is a meson whereas the neutron is a baryon). [1] *(3 marks)*

8 a Q = 0, S = −2 [1]   b Q = 0, S = −1 [1]
c Q = +1, S = +1 [1]   d Q = +1, S = +1 [1] *(4 marks)*

Chapter 3

1 a The photoelectric effect is the emission of electrons from a metal surface when electromagnetic radiation is directed at the metal surface. [1] *(1 mark)*

b Electromagnetic radiation of a certain frequency f is composed of photons each of energy $E = hf$. [1] When photons of frequency f are directed at the surface, conduction electrons in the metal near the surface may each absorb a photon. [1] An electron that absorbs a photon therefore gains energy hf when it absorbs a photon. [1] An electron cannot escape if its energy is less than the work function ϕ of the metal. Therefore, electrons cannot escape if the frequency of the incident radiation is less than $\frac{\phi}{h}$. [1] *(4 marks)*

2 a Threshold frequency is the minimum frequency of incident electromagnetic radiation on a particular metal surface that will cause emission of electrons from the surface. [1] *(1 mark)*

b The work function of the metal $\phi = hf_{min} = 6.63 \times 10^{-34} \times 2.9 \times 10^{14} = 1.92 \times 10^{-19}$ J [1]

Photon energy $= hf = \frac{hc}{\lambda}$

$= \frac{6.63 \times 10^{-34} \times 3.00 \times 10^8}{560 \times 10^{-9}} = 3.54 \times 10^{-19}$ J [1]

Maximum kinetic energy of a photoelectron $= hf - \phi = 3.54 \times 10^{-19} - 1.92 \times 10^{-19} = 1.6 \times 10^{-19}$ J (to 2 significant figures). [1] *(3 marks)*

Note: The final answer is given to 2 significant figures (sf) because the work function is only given to 2 sf. All the working up to the final answer has been done to 3 sf to avoid intermediate rounding-off causing an arithmetical error in the final answer.

c Graph is a straight line with a positive gradient. [1] Line intercepts the x-axis at f_{min}. [1] *(2 marks)*

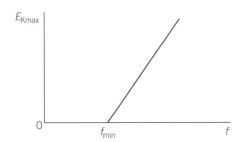

▲ Figure 1

3 a The threshold frequency $f_{min} = \dfrac{\phi}{h} = \dfrac{0.85\,eV}{h}$

$= \dfrac{0.85 \times 1.60 \times 10^{-19}\,J}{6.63 \times 10^{-34}}\,Js$ [1]

$= 2.05 \times 10^{14}\,Hz = 2.1 \times 10^{14}\,Hz$ (to 2 significant figures). [1]

(2 marks)

b Photon energy $= hf = \dfrac{hc}{\lambda}$

$= \dfrac{6.63 \times 10^{-34} \times 3.00 \times 10^8}{420 \times 10^{-9}} = 4.74 \times 10^{-19}\,J$ [1]

Work function $\phi = 0.85\,eV = 0.85 \times 1.6 \times 10^{-19}\,J$

$= 1.36 \times 10^{-19}\,J$ [1]

Maximum kinetic energy of a photoelectron

$= hf - \phi = 4.74 \times 10^{-19} - 1.36 \times 10^{-19}$

$= 3.38 \times 10^{-19}\,J$ [1]   *(3 marks)*

4 The frequency of blue light is greater than the frequency of red light. So a blue light photon has more energy than a red light photon. [1] In this example, the energy of a blue light photon is greater than the work function of the metal whereas the energy of a red light photon is less than the work function [1]. Therefore, when blue light is used, a blue light photon can give a conduction electron enough energy to overcome the work function of the metal and escape from its surface whereas a red light photon does not have enough energy to overcome the work function and escape. [1]   *(3 marks)*

5 a See Figure 1. Four energy levels shown and labelled correctly [1] gaps approx 1, 3, 5 [1]

(2 marks)

```
                                    4.6 eV
          d   e   f

                                    2.1 eV

      b   c
                                    0.5 eV
  a
                                    0
```

▲ Figure 1

b 6 photon energies *a–f* as follows, shown by the downward arrows on Figure 1 are possible:

a 0.5 eV, [1] *b* 1.6 eV (= 2.1 − 0.5 eV), *c* 2.1 eV, [1] *d* 2.5 eV (= 4.6 − 2.1 eV), *e* 4.1 eV (= 4.6 − 0.5 eV), *f* = 4.6 eV. [1]   *(3 marks)*

6 The wavelengths of the lines of a line spectrum of an element are characteristic of the atoms of that element because the energy levels of the atoms of an element are unique to that element. [1] When an electron in an atom moves from an energy level to a lower energy level, it releases a photon of energy equal to the energy difference between the two energy levels. [1] The photons that produce each line all have the same energy, which is different from the energy of the photons that produce any other line. The photon energies are therefore characteristic of the atom. [1] The wavelength of each photon depends on its energy so the photon wavelengths are characteristic of the atoms of the element. Therefore, the line spectrum is characteristic of the element. [1]   *(4 marks)*

7 a i particle nature of light

ii the wave nature of matter [1]   *(1 mark)*

b i momentum $p = \dfrac{h}{\lambda} = \dfrac{6.63 \times 10^{-34}}{500 \times 10^{-9}}$

$= 1.33 \times 10^{-27}\,kg\,m\,s^{-1}$ [1]

velocity $= \dfrac{momentum}{mass} = \dfrac{1.33 \times 10^{-27}}{9.11 \times 10^{-31}}$

$= 1460\,m\,s^{-1}$ [1]

ii momentum for the same de Broglie wavelength is the same, i.e.,

$1.33 \times 10^{-27}\,kg\,m\,s^{-1}$ [1]

velocity $= \dfrac{momentum}{mass} = \dfrac{1.33 \times 10^{-27}}{1.67 \times 10^{-27}}$

$= 0.80\,m\,s^{-1}$ [1]   *(4 marks)*

Chapter 4

1 a The vibrations of a polarised wave are perpendicular to the direction of travel of the wave along a line in one direction only. A transverse wave can therefore be polarised because its vibrations are perpendicular to the direction of travel of the wave. [1] A longitudinal wave cannot be polarised because its vibrations are along the direction of travel of the wave. [1]   *(2 marks)*

b The intensity of the light transmitted through both filters decreases as the filter is turned and becomes zero when the filter has been turned through 90°. [1] When the filter is turned through a further angle of 90°, the intensity increases from zero and then reaches a maximum again after being turned through a total angle of 180° from its initial position. [1] When the filter is turned through a further angle of 180°, the intensity varies in exactly the same way as it did when it was turned through 180° from its initial position. [1]   *(3 marks)*

2 a The microwaves reflected by the metal plate pass through the microwaves from the transmitter and so a stationary wave pattern is formed. [1] The maxima and minima are respectively the positions of the antinodes and nodes of the stationary wave pattern. [1]

Each node is where the two sets of waves always have equal and opposite displacements so the resultant amplitude is always zero. [1] Each antinode is a position where the wave peaks always arrive at the same time so the resultant amplitude is a maximum. [1]   (4 marks)

b There are 4 nodes between the 1st and last node so the distance between the 1st and last node is equal to 5 node-to-node spacings. [1] The distance between adjacent nodes is therefore $\frac{82}{5}$ mm which is 16.4 mm. [1] Adjacent nodes are half a wavelength apart therefore the wavelength is equal to 32.8 mm = 2 × 16.4 mm. [1]   (3 marks)

3 a i See Figure 1. Four equally-spaced loops [1] Antinodes (A) and nodes (N) correctly labelled [1]   (2 marks)

▲ Figure 1

ii X and Y have the same amplitude [1], and their phase difference is π radians (or 180°) [1]   (2 marks)

b i λ = 400 mm [1] as node-node spacing = 0.5λ = 200 mm [1]   (2 marks)

ii The frequency of the first harmonic $f_1 = \frac{1}{4}$ × 200 Hz [1] (because the stationary wave pattern is the 4th harmonic).

Rearranging $f_1 = \frac{1}{2L}\sqrt{\frac{T}{\mu}}$ gives $T = (2Lf_1)^2 \mu$ [1]

= $(2 \times 0.800 \times 50)^2 \times 7.4 \times 10^{-4}$ [1] = 4.7 N [1]
    (4 marks)

4 a $\lambda = \frac{c}{f} = \frac{1500}{1.6 \times 10^6} = 9.4 \times 10^{-4}$ m [1]   (1 mark)

b i Period $T = \frac{1}{f} = \frac{1}{1.6 \times 10^6} = 6.25 \times 10^{-7}$ s = 0.625 µs [1]

Therefore, the number of waves in 10 µs = $\frac{10\,\mu s}{0.625\,\mu s}$ = 16 [1]   (2 marks)

ii Any two of the following reasons: [2]

1 The waves spread out as they travel away from the probe and therefore their intensity decreases so their amplitude decreases.

2 The waves are only partially reflected as some of the wave energy is transmitted at each boundary. So the reflected waves are less intense and therefore have a smaller amplitude.

3 Some of the wave energy is absorbed by the body tissue as the waves travel through it. So the amplitude of the waves gradually decreases as they travel through the tissue.
    (2 marks)

5 a Coherent sources emit waves of the same frequency with a constant phase difference. [1]   (1 mark)

b At certain points along XY, the waves from the two sources arrive out of phase by 180°. [1] Therefore, the waves partly or totally cancel each other out. So the amplitude is a minimum at these positions. [1]   (2 marks)

6 a i The pipe length $L = \frac{1}{4}\lambda$ so $\lambda = 4L = 4 \times 380$ mm = 1520 mm. [1] The speed of sound in the pipe = $f\lambda$ = 225 Hz × 1.52 m = 342 m s^{-1} [1]

ii The amplitude decreases gradually from the open end to the closed end where it is zero. [1]
    (3 marks)

b At 675 Hz, the frequency is 3 times the first harmonic frequency of 225 Hz. The 3rd harmonic stationary wave pattern is set up when the pipe length = $\frac{3}{4}\lambda$ [1] and there is an extra node and antinode in the pipe compared with the 1st harmonic, as shown in Figure 2. [1]   (2 marks)

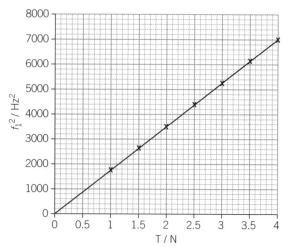

▲ Figure 2

7 a Suspend a mass of known weight W from the end of the string supported by the pulley. The tension T in the string is equal to W. [1] Increase the signal generator frequency from zero until the string vibrates with a node at either end and a single antinode in the middle. [1] This is the 1st harmonic mode of vibration. [1] Measure the signal generator frequency and the tension and record the measurements in a table. [1] Repeat the procedure for a range of different weights. [1] (max 4)

b i Square both sides of $f_1 = \frac{1}{2L}\sqrt{\frac{T}{\mu}}$ to give

$f_1^2 = \frac{1}{4L^2} \times \frac{T}{\mu} = \frac{1}{4\mu L^2}T$ [1]. Therefore, a graph of $y = f_1^2$ against $x = T$ should be a straight line through the origin of the form $y = mx$ where the gradient $m = \frac{1}{4\mu L^2}$ [1]   (2 marks)

ii Graph: suitable scales on both axes [1], points plotted correctly [1], best fit line [1]

Large gradient triangle and correct use [1] to give gradient in the range 1700 to 1850 (Hz^2 N^{-1}) [1]

Calculation of μ from $m = \frac{1}{4\mu L^2}$ to give

$\mu = \frac{1}{4\mu L^2} = \frac{1}{4 \times 1770 \times (0.64)^2}$ = 3.4 to 3.5 × 10^{-4} kg m^{-1} [1]   (6 marks)

▲ Figure 1

Chapter 5

1 **a i** See Figure 1 [1]   (*1 mark*)

ii $n_1 \sin\theta_1 = n_2 \sin\theta_2$ gives $1 \sin 35 = 1.55 \sin r$. [1]

Therefore, $\sin r = \dfrac{1\sin 35}{1.55} = 0.370$ so $r = 21.7°$ [1]
                                                            (*2 marks*)

b i Consider angles *a* and *b* in Figure 1:

$a = 90 - 21.7 = 68.3°$; $a + b + 60° = 180°$
therefore $b = 120 - a = 51.7°$

So the angle of incidence where the ray leaves the prism = $90 - 51.7 = 38.3°$ [1]

ii Applying $n_1 \sin\theta_1 = n_2 \sin\theta_2$ at the point where the ray leaves the prism gives

$1.55 \sin 38.3 = 1\sin r$. [1] Therefore, $\sin r = 1.55 \times \sin 38.3 = 0.961$ so $r = 73.9°$ [1]   (*3 marks*)

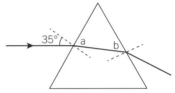

▲ **Figure 1**

2 **a** The core of a communications fibre needs to be narrow in order to ensure the light stays as near as possible to the axis of the fibres. [1] This is because non-axial rays take longer than axial rays to travel along the fibres and if this difference is too great, the pulses of light lengthen and merge before they reach the detector. [1]   (*2 marks*)

b i $n_1 \sin\theta_1 = n_2 \sin\theta_2$ at the air–core boundary
$1.55 \sin c = 1.45 \sin 90.0$. [1]

Therefore, $\sin c = 1.45 \div 1.55 = 0.935$
so $c = 69.3°$ [1]   (*2 marks*)

ii If the cladding was absent, light would pass between fibres where they are in contact instead of undergoing total internal reflection. [1] The cladding ensures total internal reflection occurs at the core–cladding boundary which means the light signals in each fibre are secure as they cannot pass between fibres that are in contact. [1]   (*2 marks*)

3 **a i** Applying $n_1 \sin\theta_1 = n_2 \sin\theta_2$ to the red component gives $1.52 \sin 40.0 = 1 \sin r$. [1] Therefore, $\sin r = 1.52 \times \sin 40.0 = 0.9770$ so $r = 77.7°$ [1]

ii Applying $n_1 \sin\theta_1 = n_2 \sin\theta_2$ to the blue component gives $1.55 \sin 40.0 = 1\sin r$. [1] Therefore, $\sin r = 1.55 \times \sin 40.0 = 0.9963$ so $r = 85.1°$ [1]   (*4 marks*)

b The angle between the red and blue components = $85.1 - 77.7 = 7.4°$ [1]   (*1 mark*)

4 **a** The two slits emit light waves of the same frequency and with a constant phase difference. [1] Therefore, at any position on the screen where the waves overlap, the waves from the two slits have a constant phase difference that depends on the difference in the distance from that position to each slit (i.e., the path difference). [1]

At positions where the path difference is a whole number of wavelengths, the waves arrive in phase so they reinforce and a bright fringe is seen at that position. [1] Midway between the bright fringes, the path difference is a whole number of wavelengths plus one-half wavelength. [1] The waves therefore arrive out of phase by 180° so they cancel and a dark fringe is seen midway between adjacent bright fringes. [1]   (*4 marks*)

b 6 fringes mean that there are 5 fringe spacings in 65 mm so the fringe spacing between adjacent fringes $W = 65\,\text{mm} \div 5 = 13\,\text{mm}$ [1]
Using the equation $s = \dfrac{\lambda D}{W}$ gives

$\lambda = \dfrac{630 \times 10^{-9} \times 2.50}{13 \times 10^{-3}}$ [1] $= 1.2 \times 10^{-4}$ m [1]   (*3 marks*)

5 **a** See Topic 5.6, Figure 1. Your graph should show each outer bright fringe is half the width of the central bright fringe [1] which should be at least 3 times higher than the nearest bright fringe on each side of the central fringe. [1] The graph should be symmetrical about a line through the centre of the central bright fringe. [1]   (*3 marks*)

b If the slit is made wider, the bright fringes become brighter and closer together. [1] The central bright fringe is still twice the width of the other bright fringes and is still about 3 times higher than the nearest fringe on each side. [1]   (*2 marks*)

6 **a** The maximum order number = $\dfrac{d}{\lambda}$ rounded down where $d = \dfrac{1}{600}$ mm $= 1.67 \times 10^{-6}$ m. [1] Use the largest wavelength value of 445 nm to give the smallest value of $\dfrac{d}{\lambda} = \dfrac{1.67 \times 10^{-6}}{445 \times 10^{-9}} = 3.75$ m. [1]
Rounding this down gives 3 for the maximum order number. [1] Smaller wavelengths are diffracted less by the grating so the 3rd order will have the full range of wavelengths from 430 to 445 nm. [1]   (*4 marks*)

b Using the 3rd order equation $3\lambda = d \sin\theta$ for 430 nm and 445 nm in turn gives the angle of diffraction for these wavelengths.

For $\lambda = 430$ nm, $\sin\theta = \dfrac{3\lambda}{d} = \dfrac{3 \times 430 \times 10^{-9}}{1.67 \times 10^{-6}}$
$= 0.772$ [1] which gives $\theta = 50.57°$

For $\lambda = 445$ nm, $\sin\theta = \dfrac{3\lambda}{d} = \dfrac{3 \times 445 \times 10^{-9}}{1.67 \times 10^{-6}}$
$= 0.799$ [1] which gives $\theta = 53.07°$

The angular width of the 3rd order beam is therefore $53.07 - 50.57 = 2.50°$ [1]   (*3 marks*)

7 **a i** If slit S was not narrow enough, the bright fringes would be wider and the dark fringes narrower. [1] This is because adjacent strips of a wide slit would each produce diffraction patterns slightly displaced from each other. [1] The resultant pattern would therefore have reduced darkness between the bright fringes. [1] So S needs to be narrow enough to give bright and dark fringes of equal width. [1]   (*4 marks*)

ii S_1 and S_2 are coherent sources because each slit emits a wavefront every time a wavefront from S reaches S_1 and S_2. [1] So S_1 and S_2 emit waves with a constant phase difference and are therefore coherent sources. [1]   (*2 marks*)

b The intensity of the fainter fringes has been suppressed by single slit diffraction at S_1 and S_2 [1] because these fringes are at a single slit diffraction minimum. [1] Adjacent fringes further from the centre are brighter because they are not at a single slit diffraction minimum. [1]   (*3 marks*)

8 a $d = \dfrac{1}{600}$ mm $= 1.67 \times 10^{-6}$ m. [1] Using the 2nd order equation $2\lambda = d\sin\theta$ gives

$\lambda = \dfrac{d\sin\theta}{2} = \dfrac{1.67 \times 10^{-6} \times \sin 40.0}{2} = 5.37 \times 10^{-7}$ m. [1]

(*3 marks*)

b The wavelength of the light in glass

$= \dfrac{\text{wavelength in air}}{\text{refractive index of glass}} = \dfrac{5.37 \times 10^{-7}}{1.50} = 3.58 \times 10^{-7}$ m.

[1] Using the 2nd order equation gives

$\sin\theta = \dfrac{2\lambda}{d} = \dfrac{2 \times 3.58 \times 10^{-7}}{1.67 \times 10^{-6}} = 0.429$ [1] which gives

$\theta = 25.4°$ [1]   (*4 marks*)

Chapter 6

1 a See Figure 1 [1] $a = 56°$ [1] $b = 30°$ [1]. Note the construction arcs drawn using the same scale as the 6 N vector.   (*3 marks*)

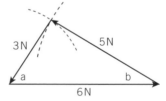

▲ **Figure 1**

b 5 N in the opposite direction to the 5 N vector in Figure 1. [1]   (*1 mark*)

2 The tension in each section of the wire is the same because their horizontal components are equal and opposite to each other [1] and the angle between each wire and the horizontal is the same (i.e., $T_1\cos 15 = T_2\cos 15$) [1].

The sum of their vertical components is equal and opposite to the weight.

(i.e., $T_1\sin 15 + T_2\sin 15 = W$). [1] Hence, $2\,T_1\sin 15 = W$ so $T_1 = T_2 = \dfrac{W}{2\sin 15} = \dfrac{6.3}{2\sin 15} = 12.2$ N [1]

(*4 marks*)

3 a See the parallelogram of forces in Figure 2 [1]. The angle between the resultant and the 7.2 kN force is 24° [1] and is 21° between the 8.0 kN force and the resultant. [1]   (*3 marks*)

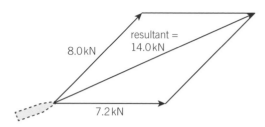

▲ **Figure 2**

b i The resultant is 1.75 times longer than the 8.0 kN force [1] so its magnitude is 14 kN [1] ($= 1.75 \times 8.0$ kN).

ii The drag force is 14.0 kN in the opposite direction to the resultant force. [1]   (*3 marks*)

4 a See Figure 3: resolve T into a horizontal component $T\cos 60$ and a vertical component $T\sin 60$. Taking moments about P gives $T\sin 60 \times (0.950 - 0.350)$ [1] $= W \times (0.500 - 0.350)$ [1]

Therefore, $T = \dfrac{1.20 \times 0.150}{0.600\sin 60} = 0.346 = 0.35$ N [1] to 2 significant figures.   (*3 marks*)

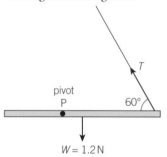

▲ **Figure 3**

b The horizontal component of the support force $S_x = T\cos 60 = 0.346 \times \cos 60 = 0.173$ N [1]

The vertical component of the support force $S_y = W - T\sin 60 = 1.2 - (0.346 \times \sin 60) = 0.900$ N [1]

The magnitude of the support force $S = (S_x^2 + S_y^2)^{1/2} = 0.92$ N [1] to 2 significant figures.

The angle of the line of action of S to the horizontal $= \arctan(S_y \div S_x) = 79°$ [1]

Note the line of action of S should pass through the point where the lines of action of W and T intersect which will be the position on the string vertically above the centre of mass.   (*4 marks*)

5 a See Figure 4 where C is the centre of mass of the plank. [1]   (*1 mark*)

b Distances XP $= 1.50 - 0.20 = 1.30$ m, PC $= 0.50$ m, and CY $= 2.00 - 0.20 = 1.80$ m

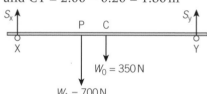

▲ **Figure 4**

Taking moments about Y to find the support force S_x at X gives:

Anticlockwise moment = $(700 \times PY) + (350 \times CY)$
$= (700 \times 2.30) + (350 \times 1.80) = 2240\,\text{N m}$[1]

Clockwise moment = $S_x \times XY = 3.60\,S_x$ [1]

Therefore, $3.60\,S_x = 2240$ hence $S_x = \dfrac{2240}{3.60} = 622\,\text{N}$ [1]

Since $S_x + S_y$ = the total weight = 700 + 350 = 1050 N, then $S_y = 1050 - 622 = 428\,\text{N}$ [1]

(4 marks)

6 a The total weight supported by the curtain brackets is 108 N (= 22 N + 86 N). [1] The brackets are equidistant from the centre of mass of the pole which is at the centre of the pole. The curtains are symmetrical about the centre of mass. [1] Therefore, each bracket supports half of the total weight which means the support force of each bracket on the pole is 54 N vertically upwards. [1] *(3 marks)*

b The support force S_x at X increases as the centre of mass of the curtain moves towards X. [1] When the curtain is at X, the bracket at X supports the weight of the curtain that has been pulled back and also the force of the pole on X due to the other curtain and the weight of the pole. [1] *(2 marks)*

c See Figure 5: M is the centre of mass of the plank. W_1 and W_2 represent the weight of each curtain (= 43 N each).

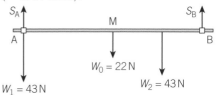

▲ **Figure 5**

The line of action of W_1 passes through bracket A and the line of action of W_2 passes through the point on the curtain pole midway between M and B.

To determine the support force S_A at A, take moments about B:

Distance AB = $2.80 - (2 \times 0.05) = 2.70\,\text{m}$, AM = MB = $\dfrac{2.80}{2} - 0.05 = 1.35\,\text{m}$.

The perpendicular distance from B to the line of action of $W_2 = \frac{1}{2}$ MB = 0.675 m.

Taking moments about B to find the support force S_A at A gives:

Clockwise moment = $S_A \times AB = 2.70\,S_A$ [1]

Anticlockwise moment = $(43 \times AB) + (22 \times MB) + (43 \times \frac{1}{2}\,MB)$
$= (43 \times 2.70) + (22 \times 2.30) + (43 \times 0.675)$ [1]
$= 196\,\text{N m}$ [1]

Therefore, $2.70\,S_A = 196$ [1] hence $S_A = \dfrac{196}{2.70} = 73\,\text{N}$ [1] (to 2 significant figures)

(5 marks)

7 Figure 6 shows the forces acting at P.

Resolving the forces vertically and horizontally gives:

a Horizontally: $T_1 \cos 20 = T_2 \cos 15$ [1] hence $T_1 = \dfrac{\cos 15}{\cos 20}\,T_2 = 1.03\,T_2$ [1] *(2 marks)*

b Vertically: $T_1 \sin 20 + T_2 \sin 15 = W$ hence $(1.03 \sin 20 + \sin 15)\,T_2 = 30$ [1]

Therefore, $0.611\,T_2 = 30$ [1] so $T_2 = \dfrac{30}{0.611} = 49.1\,\text{N}$ [1] to 2 significant figures.

Since $T_1 = 1.03\,T_2$ then $T_2 = 1.03 \times 49.1 = 51\,\text{N}$ [1] to 2 significant figures. *(4 marks)*

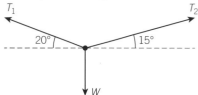

▲ **Figure 6**

8 a Applying the principle of moments about the wheel axis when a horizontal force F is applied to the handle gives $F \times 0.85\,\text{m} = W \times 0.16\,\text{m}$ [1] hence $F = \dfrac{0.16}{0.85} \times 220\,\text{N} = 41\,\text{N}$ [1] *(2 marks)*

b The suitcase is unstable in this position because its weight will have a non-zero moment about the wheel axis if it is displaced slightly from this position [1]. As a result, it will topple forwards or backwards if it is displaced slightly from this position. [1] *(2 marks)*

c An upward force is necessary to provide a moment in the opposite direction to the moment of the weight about the wheel axis [1] which acts in a clockwise direction. The force on the handle must therefore be upwards to provide an anticlockwise moment about the wheel axis. [1] *(2 marks)*

Chapter 7

1 $u = 98\,\text{km h}^{-1} = 27.2\,\text{m s}^{-1}, v = 0, s = 73\,\text{m}$

a To find a, rearrange $v^2 = u^2 + 2as$ to give $a = \dfrac{v^2 - u^2}{2s} = \dfrac{0 - 27.2^2}{2 \times 73}$ [1] $= -5.07\,\text{m s}^{-2} = 5.1\,\text{m s}^{-2}$ [1] to 2 significant figures *(2 marks)*

b To find t, rearrange $v = u + at$ to give $t = \dfrac{v - u}{a} = \dfrac{0 - 27.2}{-5.07}$ [1] $= 5.4\,\text{s}$ [1] *(2 marks)*

2 a The gradient and therefore the acceleration is greatest in magnitude at $t = 30\,\text{s}$. [1]

Drawing a tangent at this point gives the maximum acceleration = gradient of the tangent. [1]
$= \dfrac{0 - 80}{55 - 4} = -1.54\,\text{m s}^{-2} = -3.25\,\text{m s}^{-2}$ [1] *(3 marks)*

b Distance travelled = area under the line (= 89 + 76 + 59 + 37 + 17 + 4) = 282 small squares where 25 small squares = 1 large square = 200 m. [1] Therefore, the distance travelled = (282 ÷ 25) × 200 m = 175 N. [1] *(2 marks)*

3 a i For the first 1.2 s, $u = 0$, $t = 1.20$ s, $v = 9.60\,\text{m s}^{-1}$

Rearranging $v = u + at$ gives $a = \dfrac{v - u}{t} = \dfrac{9.60 - 0}{1.20}$

$= 8.00\,\text{m s}^{-2}$ [1]

(1 mark)

ii Distance travelled in first 1.20 s, $s = ut + \dfrac{1}{2}at^2$

$= 0 + \dfrac{1}{2} \times 8.00 \times 1.20^2 = 5.76\,\text{m}$ [1]

Therefore, the distance travelled at $9.6\,\text{m s}^{-1}$

$= 100\,\text{m} - 5.76\,\text{m} = 94.24\,\text{m}$

Time taken to run this distance $= \dfrac{\text{distance}}{\text{speed}}$

$= \dfrac{94.24}{9.6} = 9.82\,\text{s}$ [1]

Therefore, the total time taken $= 1.20 + 9.82$

$= 11.02\,\text{s}$ [1]

(3 marks)

b Let V represent Y's maximum speed.

The time taken at constant speed $V = 11.02 - 1.30$

$= 9.72\,\text{s}$

So the distance moved at constant speed

$V = \text{speed} \times \text{time} = 9.72V$ [1]

The distance moved by Y in 1.3 s is given by

$s = \dfrac{u + v}{2}t = \dfrac{0 - V}{2} \times 1.30 = 0.65V$ [1]

Therefore, the total distance moved by

$Y = 9.72V + 0.65V = 10.37V = 100\,\text{m}$

Hence, $V = \dfrac{100}{10.37} = 9.64\,\text{m s}^{-1}$ [1]

(3 marks)

4 a $u = 1.80\,\text{m s}^{-1}$, $v = 0$, $t = 12.0\,\text{s}$

Its rate of change of velocity $= a = \dfrac{v - u}{t} = \dfrac{0 - 1.80}{12.0}$

[1] $= -0.150\,\text{m s}^{-2}$ [1]

(2 marks)

b $u = 1.80\,\text{m s}^{-1}$, $a = -0.150\,\text{m s}^{-2}$, $t = 16.0\,\text{s}$

i To calculate v, using $v = u + at$ gives

$v = 1.80 + (-0.150 \times 16.0)$ [1] $= -0.60\,\text{m s}^{-1}$

The negative sign means the wagon is moving down the incline. [1]

(2 marks)

ii To calculate s, using $s = ut + \dfrac{1}{2}at^2$ gives

$s = (1.80 \times 16.0) + (0.5 \times -0.150 \times 16.0^2)$

$= 9.60\,\text{m}$ [1]

The position of the wagon is therefore 9.60 m from the bottom of the incline. [1]   *(2 marks)*

5 a i $u = 0$, $t = 20\,\text{s}$, $v = 8.0\,\text{m s}^{-1}$

Therefore, $s = \dfrac{u + v}{2}t = \dfrac{0 + 8.0}{2} \times 20 = 80\,\text{m}$ [1]

(1 mark)

ii $u = 8.0\,\text{m s}^{-1}$, $s = 20.0\,\text{m}$, $v = 0$

Rearrange $s = \dfrac{u + v}{2}t$ to give $t = \dfrac{2s}{u + v} = \dfrac{2 \times 20.0}{0 + 8.0}$

$t = 5.0\,\text{s}$ [1]

(1 mark)

b See Figure 1: PQR correct [1] RS correct [1] *(2 marks)*

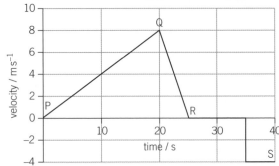

▲ **Figure 1**

c i $u = 0$, $v = 8.0\,\text{m s}^{-1}$, $t = 5.0\,\text{s}$

acceleration $a = \dfrac{v - u}{t} = \dfrac{8.0 - 0}{5.0} = 1.6\,\text{m s}^{-2}$ [1]

(1 mark)

ii Displacement = area between v–t line and axis.

Displacement PR $= \dfrac{1}{2} \times 8.0\,\text{m s}^{-1} \times 25.0\,\text{s}$

$= 100\,\text{m}$ [1]

Displacement RS $= -4.0\,\text{m s}^{-1} \times 5.0\,\text{s} = -20\,\text{m}$

Therefore, displacement PS $= 100 - 20 = 80\,\text{m}$ [1]

(2 marks)

6 a Vertical motion $u = 0$, $a = -9.81\,\text{m s}^{-2}$, $t = 2.0\,\text{s}$

Vertical displacement $y = \dfrac{1}{2}a_y t^2$ gives

$y = \dfrac{1}{2} \times -9.81 \times 2.0^2$ [1] $= -19.6\,\text{m}$ [1]

Horizontal motion $u = 14\,\text{m s}^{-1}$, $a_x = 0$, $t = 2.0\,\text{s}$

Horizontal displacement $x = u_x t = 14 \times 2.0 = 28\,\text{m}$ [1]

Distance $s = \sqrt{x^2 + y^2} = \sqrt{28^2 + (-19.6)^2} = 34\,\text{m}$ [1]

to 2 significant figures

(4 marks)

b The stone would have experienced a drag force due to air resistance. [1] The drag force increases with speed. [1] The effect of drag on the vertical component of velocity would be to slow its descent and increase the time taken to fall to the ground. [1]

Compared with the motion without air resistance, the horizontal component of velocity is reduced by the drag force more than the time to fall through a given vertical distance is increased. [1] So the effect of air resistance is to reduce the horizontal component of displacement to the point where the stone hit the water. [1]   *(5 marks)*

7 a Horizontal component $u_x = 2.60 \cos 30.0$

$= 2.25\,\text{m s}^{-1}$ [1]

Vertical component $u_y = 2.60 \sin 30.0 = 1.30\,\text{m s}^{-1}$ [1]

(2 marks)

b Consider the vertical motion: $u_y = 1.30\,\text{m s}^{-1}$, $t = 3.90\,\text{s}$, $a_y = -9.81\,\text{m s}^{-2}$ (– for downwards)

i To calculate the vertical component of displacement, y_y, using $y = u_y t + \dfrac{1}{2}a_y t^2$ gives

$y = (1.30 \times 3.90) + (\dfrac{1}{2} \times -9.81 \times 3.90^2)$ [1]

$= -69.5\,\text{m}$.

Therefore, the vertical distance fallen $= 69.5\,\text{m}$. [1]   *(2 marks)*

ii Consider the horizontal motion:

$u_x = 2.25\,\text{m s}^{-1}$, $a_x = 0$, $t = 3.90\,\text{s}$

Therefore, the horizontal displacement $x = u_x t$

$= 2.25 \times 3.90 = 8.78\,\text{m}$ [1]   *(1 mark)*

8 a i Horizontal motion: $a_x = 0$, $x = 85.5\,\text{m}$,

$t = 4.24$ s, $u_x = ?$

Horizontal displacement $x = u_x t$ therefore the horizontal component of the initial velocity

$u_x = \dfrac{x}{t} = \dfrac{85.5}{4.24} = 20.2\,\text{m s}^{-1}$ [1]   *(1 mark)*

ii Vertical motion: $t = 4.24\,s$, $y = -1.80\,\text{m}$,

$a_y = -9.81\,\text{m s}^{-2}$ (– for downwards), $u_y = ?$

To calculate the vertical component of the initial velocity, rearrange $y = u_y t + \dfrac{1}{2}a_y t^2$ to give

$u_y t = y - \frac{1}{2} a_y t^2 = -1.80 - (\frac{1}{2} \times -9.81 \times 4.24^2)$ [1]

$= 86.4\,\text{m}$ [1]

Therefore, $u_y t = 86.4\,\text{m}$ hence $u_y = \frac{86.4}{t} = \frac{86.4}{4.24}$

$= 20.4\,\text{m s}^{-1}$ [1]　　　　(3 marks)

b　i Initial speed of javelin $= \sqrt{u_x^2 + u_y^2} = \sqrt{20.2^2 + 20.4^2}$

$= 28.7\,\text{m s}^{-1}$ [1]　　　　(1 mark)

ii $\tan\theta = \frac{u_y}{u_x} = \frac{20.4}{20.2} = 1.01$ therefore $\theta = 45.3°$ [1]

(1 mark)

Chapter 8

1　a　i $u = 0, v = 8.8\,\text{m s}^{-1}, t = 40\,\text{s}, m = 1800\,\text{kg}$

acceleration $a = \frac{v-u}{t} = \frac{8.8-0}{40} = 0.22\,\text{m s}^{-2}$ [1]

force $F = ma = 1800 \times 0.22 = 400\,\text{N}$ [1]

ii $a = 0.22\,\text{m s}^{-2}, m = 300\,\text{kg}$

tension $T = ma = 300 \times 0.22 = 66\,\text{N}$ [1] (3 marks)

b force $= 396\,\text{N}$

acceleration $= \frac{\text{engine force}}{\text{mass}} = \frac{396}{1500} = 0.26\,\text{m s}^{-2}$ [1]

$u = 0, v = 8.8\,\text{m s}^{-1}, a = 0.26\,\text{m s}^{-2}$

To calculate t, rearrange $v = u + at$ to give $t = \frac{v-u}{a}$

$= \frac{8.8-0}{0.26} = 33.8\,\text{s} = 34\,\text{s}$ to 2 significant figures. [1]

(2 marks)

2　a $u = 0, v = 8.3\,\text{m s}^{-1}, s = 150\,\text{m}, m = 61\,\text{kg}$

To calculate s, rearrange $v^2 = u^2 + 2as$ to give

$a = \frac{v^2 - u^2}{2s} = \frac{8.3^2 - 0}{2 \times 74}$ [1] $= 0.465\,\text{m s}^{-2}$ [1]

resultant force = mass × acceleration

$= 61 \times 0.465 = 28\,\text{N}$ [1]　　　　(3 marks)

b Component of weight acting down the slope

$= mg\sin5°$ [1] $= 61 \times 9.81 \times \sin5.0 = 52.2\,\text{N}$ [1]

resultant force $= 52.2 + 28.3 = 80.5\,\text{N}$ [1]

acceleration $= \frac{\text{resultant force}}{\text{mass}} = \frac{80.5}{61}$

$= 1.3\,\text{m s}^{-2}$ [1]　　　　(4 marks)

3　a maximum acceleration = initial gradient

(= gradient of tangent at t = 0) [1]

$= \frac{26-0}{5.0} = 5.2\,\text{m s}^{-2}$ [1]　　　　(2 marks)

b　i Resultant force = driving force − drag force and the drag force is zero at $t = 0$. [1]

So at $t = 0$, the driving force = resultant force = mass × acceleration $= 1200 \times 5.2 = 6200\,\text{N}$ [1]

(4 marks)

ii At constant velocity, the resultant force is zero as the acceleration is zero. [1] Therefore, the drag force at this velocity is equal to the driving force. [1] So the drag force is 650 N in the opposite direction to the direction of motion of the car. [1]　　　(3 marks)

4　a　i $u = 18\,\text{m s}^{-1}, v = 0.5\,\text{m s}^{-1}, s = 28\,\text{m}, m = 750\,\text{kg}$

To calculate a, rearrange $v^2 = u^2 + 2as$ to give

$a = \frac{v^2 - u^2}{2s} = \frac{0.5^2 - 18^2}{2 \times 28}$ [1] $= -5.8\,\text{m s}^{-2}$ (2 s.f.) [1]

ii resultant force = mass × acceleration

$= 750 \times -5.78 = -4300\,\text{N}$

The resultant force is 4300 N acting down the slope. [1]　　　　(3 marks)

b　i Component of weight acting down the slope $= mg\sin32°$ [1] $= 750 \times 9.81 \times \sin32$ $= 3900\,\text{N}$ [1]

ii Resultant force down slope = component of weight down the slope + average resistive force. Average resistive force = resultant force − component of weight [1] $= 4330 - 3900$ $= 430\,\text{N}$. [1]　　　　(4 marks)

5　a The acceleration changes rapidly from zero to a large negative value when the parachute opens [1] and a large upwards force acts on the parachute due to the increase of air resistance when it opens [1]. The acceleration decreases as the parachute speed decreases and the resistive force on it decreases and therefore the upwards force on it gradually decreases [1] until the resistive force is equal to the weight. [1] The resultant force and the acceleration are then zero and the velocity hence the speed is constant. [1] (max 4)

b　i See Figure 1: steeper deceleration [1] lower final speed [1]

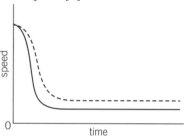

▲ Figure 1

ii The deceleration would be greater because the mass would be less and the initial upward force due to the parachute would be the same. [1] The constant speed at which the upward force is equal and opposite to the weight would be less so the flat section of the graph would be lower. [1]　　　　(4 marks)

6　a　i The overall resultant force is the difference in the weights which is 40 N (= 290 N − 250 N). [1]

Mass of the load $m_1 = \frac{250}{g} = 25.5\,\text{kg}$: mass

of the counterweight $m_2 = \frac{290}{g} = 29.6\,\text{kg}$ [1]

Therefore, the acceleration $a = \frac{\text{resultant force}}{\text{total mass}}$

$= \frac{40}{29.6 + 25.5}$ [1] $= 0.73\,\text{m s}^{-2}$ [1]　　　(4 marks)

ii Considering the forces acting on the load, the resultant force = load weight − cable tension T [1]

Therefore, $T = m_1 g - m_1 a$ [1] $= m_1(g - a)$ $= 29.6 \times (9.81 - 0.73) = 270\,\text{N}$ [1]

(or using the counterweight equation above $T = m_2 a + m_2 g = m_2(a + g)$ [1] $= 25.5 \times (9.81 + 0.73) = 270\,\text{N}$ (2 s.f.) [1])

(2 marks)

b Support force on the pulley $= 2T + W_0$ [1] where $W_0 =$ the weight of the pulley and the cable.

Therefore, the support force = $(2 \times 269) + 55$
$= 590\,\text{N}$ [1]   (*2 marks*)

7 a i $u = 30\,\text{m s}^{-1}, v = 0, s = 0.80\,\text{m}$,
impact time $t = \dfrac{2s}{u+v} = \dfrac{2 \times 0.80}{30+0} = 0.053\,\text{s}$ [1]

ii Acceleration $a = \dfrac{v-u}{t} = \dfrac{0-30}{0.0533} = -563\,\text{m s}^{-2}$ [1]
impact force $F = ma = 1800 \times -563$
$= -1.0 \times 10^6\,\text{N}$ (2 s.f.) [1] (*3 marks*)

b $u = 30\,\text{m s}^{-1}, v = 0, s = 0.80 + 0.40 = 1.20\,\text{m}, m = 71\,\text{kg}$
To calculate s, rearrange $v^2 = u^2 + 2as$ to give
$a = \dfrac{v^2 - u^2}{2s} = \dfrac{0 - 30^2}{2 \times 1.20}$ [1] $= -375\,\text{m s}^{-2}$ [1]
impact force $F = ma = 71 \times -375 = -2.67 \times 10^4\,\text{N}$ [1]
(*3 marks*)

8 a i $u = 0, t = 5.6\,\text{s}, s = 15\,\text{m}$
To calculate acceleration a, rearrange
$s = ut + \dfrac{1}{2}at^2$ with $u = 0$ to give
$a = \dfrac{2s}{t^2} = \dfrac{2 \times 15}{5.6^2}$ [1] $= 0.957\,\text{m s}^{-2} = 0.96\,\text{m s}^{-2}$ to
2 significant figures [1]

ii Resultant force $F = ma = 38 \times 0.957 = 36.4\,\text{N}$
$= 36\,\text{N}$ to 2 significant figures [1]   (*3 marks*)

b i Component of weight down the slope
$= mg\sin\theta = 38 \times 9.81 \times \sin 6.1$ [1]
$= 39.6\,\text{N} = 40\,\text{N}$ to 2 significant figures [1]
(*2 marks*)

ii Resistive forces due to frictional forces at the
bearings of the skateboard wheels and air
resistance act in the opposite direction to the
direction of motion. [1]

The resultant force = the component of the
weight acting down the slope – the total
resistive force, so the resultant force is less
than the component of weight acting down
the slope. [1] The difference is the resistive
force which is therefore equal to 3.2 N
(= 39.6 N – 36.4 N). [1]   (*2 marks*)

Chapter 9

1 a For any system of interacting objects, their total
momentum remains constant, provided no external
resultant force acts on the system. [1]   (*1 mark*)

b i Let V = the velocity of the vehicles after the
collision
Total initial momentum = $(1200 \times 21.0) +$
$(1600 \times 3.0) = 30\,000\,\text{kg m s}^{-1}$ [1]
Total final momentum = $(1200 + 1600) \times V$
$= 2800V$ [1]
Therefore, $2800V = 30\,000$ which gives
$V = \dfrac{30\,000}{2800} = 11\,\text{m s}^{-1}$ (2 s.f.) [1]

ii Kinetic energy before collision =
$\left(\dfrac{1}{2} \times 1200 \times 21.0^2\right) + \left(\dfrac{1}{2} \times 1600 \times 3.0^2\right)$
$= 2.72 \times 10^5\,\text{J}$ [1]
Kinetic energy after collision $= \dfrac{1}{2} \times$
$(1200 + 1600) \times 10.7^2 = 1.61 \times 10^5\,\text{J}$ [1]

Loss of kinetic energy $= 2.72 \times 10^5\,\text{J} -$
$1.61 \times 10^5\,\text{J} = 1.1 \times 10^5\,\text{J}$ [1]   (*6 marks*)

2 a Component of momentum perpendicular to the
wall $= 0.44 \times 24 \times \cos 30 = 9.1\,\text{kg m s}^{-1}$ [1] (*1 mark*)
Change of momentum $= 2 \times 9.1\,\text{kg m s}^{-1}$
$= 18\,\text{kg m s}^{-1}$ (2 s.f.) [1]

b Force $= \dfrac{\text{change in momentum}}{\text{contact time}} = \dfrac{18.2}{0.095} = 190\,\text{N}$ [1]
(*2 marks*)

3 a An elastic collision is one in which the total
kinetic energy after the collision is equal to the
total kinetic energy before the collision.   (*1 mark*)

b i Let V = the velocity of Y after the collision
Total initial momentum = $(0.12 \times 0.58) -$
$(0.10 \times 0.50) = 1.96 \times 10^{-2}\,\text{kg m s}^{-1}$ [1]
Total final momentum = (0.12×0.15)
$+ (0.10 \times V) = 0.018 + 0.10V$ [1]
Therefore, $0.10V + 0.018 = 1.96 \times 10^{-2}$ which
gives $V = \dfrac{1.96 \times 10^{-2} - 0.018}{0.10} = 0.016\,\text{m s}^{-1}$
The velocity of Y is therefore $0.016\,\text{m s}^{-1}$ in the
same direction as X initially. [1]   (*3 marks*)

ii Kinetic energy before collision =
$\left(\dfrac{1}{2} \times 0.12 \times 0.58^2\right) + \left(\dfrac{1}{2} \times 0.10 \times -0.50^2\right)$
$= 3.27 \times 10^{-2}\,\text{J}$ [1]
Kinetic energy after collision $= \left(\dfrac{1}{2} \times 0.12 \times 0.15^2\right)$
$+ \left(\dfrac{1}{2} \times 0.10 \times 0.016^2\right) = 1.36 \times 10^{-3}\,\text{J}$ [1]
The collision is inelastic as the total final
kinetic energy is less than the total initial
kinetic energy. [1]   (*3 marks*)

4 a i Let V = velocity of A after A and B move apart
Total final momentum = $(1.20 \times V) +$
$(0.80 \times 0.15) = 1.2V + 0.12$
Total initial momentum = 0 [1]
Therefore, $1.20V + 0.12 = 0$ [1] which gives
$V = -\dfrac{0.12}{1.20} = -0.10\,\text{m s}^{-1}$
The velocity of A = $0.10\,\text{m s}^{-1}$ in the opposite
direction to that of B. [1]   (*3 marks*)

ii Kinetic energy of A $= \dfrac{1}{2} \times 1.20 \times 0.10^2$
$= 6.0 \times 10^{-3}\,\text{J}$ [1]
Kinetic energy of B $= \dfrac{1}{2} \times 0.80 \times 0.15^2$
$= 9.0 \times 10^{-3}\,\text{J}$ [1]
Total kinetic energy $= 6.0 \times 10^{-3} + 9.0 \times 10^{-3}$
$= 0.015\,\text{J}$ [1]   (*3 marks*)

b Let V = the speed of B
Kinetic energy of B $= \dfrac{1}{2} \times 0.80 \times v^2$ [1] $= 0.015\,\text{J}$
Therefore, $v^2 = \dfrac{2 \times 0.015}{0.80} = 0.0375$ so
$v = 0.19\,\text{m s}^{-1}$ [1]   (*2 marks*)

5 a [1] for both mass numbers correct, [1] for
Th atomic number correct.   (*2 marks*)
$^{210}_{83}\text{Bi} \rightarrow \,^{4}_{2}\alpha + \,^{206}_{81}\text{Th}$

b i Let V = recoil velocity of the thallium nucleus

Total final momentum = $(4u \times 1.5 \times 10^7 \text{m s}^{-1})$ + $(206u \times V)$

Total initial momentum = 0 [1]

Therefore, $(4u \times 1.5 \times 10^7 \text{m s}^{-1})$ + $(206u \times V)$ = 0 [1]

which gives $V = \dfrac{-(4u \times 1.5 \times 10^7)\text{ m s}^{-1}}{206u}$

= $-2.9 \times 10^5 \text{m s}^{-1}$

The recoil velocity of the thallium nucleus = $2.9 \times 10^5 \text{m s}^{-1}$ [1]   *(3 marks)*

ii Kinetic energy of α particle = $\dfrac{1}{2} \times 4 \times$ $6.7 \times 10^{-27} \times (1.5 \times 10^7)^2 = 3.0 \times 10^{-12}\text{J}$ [1]

Kinetic energy of thallium nucleus = $\dfrac{1}{2} \times (206$ $\times \dfrac{1}{4} \times 6.7 \times 10^{-27}) \times (2.9 \times 10^5)^2 = 1.5 \times 10^{-14}\text{J}$ [1]

Therefore, kinetic energy of the α particle as a % of the total kinetic energy

= $\dfrac{3.0 \times 10^{-12}\text{J}}{(3.0 \times 10^{-12}\text{J}) + (1.5 \times 10^{-14}\text{J})} \times 100\%$

= 99.5% [1]   *(3 marks)*

6 a The impulse (or the change of momentum) of the ball. [1]   *(1 mark)*

b i Change of momentum = area under curve = 520×10^{-3} Ns. Answer in the range 520 ± 40. [1] Answer in the range 520 ± 20. [2]   *(2 marks)*

ii Initial momentum = 0 therefore final momentum = 0.52 Ns [1]

Velocity after impact = $\dfrac{\text{change of momentum}}{\text{mass}}$

= $\dfrac{0.52\,\text{N s}}{0.056\,\text{kg}} = 9.3 \text{m s}^{-1}$ [1]   *(2 marks)*

iii Impact time ≈ 30 ms [1]

Average acceleration = $\dfrac{\text{change of velocity}}{\text{time taken}} \approx$

$\dfrac{9.3 \text{ m s}^{-1}}{30 \times 10^{-3} \text{ s}} = 310 \text{m s}^{-2}$ [1]   *(2 marks)*

7 a i $u = 0$, $s = -0.80$ m, $a = -9.81 \text{m s}^{-2}$, $v = ?$

Using $v^2 = u^2 + 2as$ gives $v^2 = 2 \times -9.81 \times -0.80$ = $15.7 \text{m}^2\text{s}^{-2}$ [1] hence $v = 4.0 \text{m s}^{-1}$ [1]

Momentum just before impact = 4000×4.0 = $1.6 \times 10^4 \text{kg m s}^{-1}$ [1]   *(3 marks)*

ii Total momentum just after impact = (4000 + 2000) $V = 6000V$ [1] where V is the velocity of the hammer and the pile just after impact.

Using the principle of conservation of momentum gives $6000 V = 1.6 \times 10^4$ [1]

Therefore, $V = \dfrac{1.6 \times 10^4}{6000} = 2.7 \text{m s}^{-1}$ [1]   *(3 marks)*

b i $u = 2.7 \text{m s}^{-1}$, $s = -0.020$ m, $v = 0$, $a = ?$

Rearranging $v^2 = u^2 + 2as$ gives

$a = \dfrac{-u^2}{2s} = \dfrac{-2.7^2}{2 \times 0.020}$ [1] = 180m s^{-2} (2 s.f.) [1]

ii The total mass decelerated by the frictional force = 6000 kg

Frictional force $F = ma = 6000 \times 180$ = 1.1×10^6 N [1]   *(3 marks)*

8 a Mass per second transferred = $\dfrac{2.7}{60} = 0.045 \text{kg s}^{-1}$ [1]

Momentum transferred per second = mass transferred per second × velocity of the water jet = 0.045×400 [1]= 18 N [1]   *(3 marks)*

b The force of the surface on the jet = momentum loss per second = 18 N [1]

The force of the jet on the surface is equal and opposite to the force of the surface on the jet so the force of the jet on the surface is 18 N. [1]   *(2 marks)*

c Area of cross-section of the jet = $\dfrac{1}{4}\pi d^2$ = $0.25 \times \pi \times (3.8 \times 10^{-4})^2 = 1.1 \times 10^{-7}\text{m}^2$ [1]

pressure = $\dfrac{\text{force}}{\text{area}} = \dfrac{18 \text{ N}}{1.1 \times 10^{-7}\text{m}^2} = 1.6 \times 10^8 \text{Pa}$ (or N m^{-2}) [1]   *(2 marks)*

Chapter 10

1 a i Loss of potential energy = $mg\Delta h = 52.0 \times 9.81$ $\times 2.50 = 1280 \text{J}$ [1]

ii Gain of kinetic energy (KE) = $\dfrac{1}{2}mv^2$ = $0.5 \times 52.0 \times 2.12^2 = 117 \text{J}$ [1]   *(2 marks)*

b Work done to overcome friction = $mg\Delta h - \dfrac{1}{2}mv^2$ [1] = $1280 - 117 = 1163 \text{J}$ [1]

Using work done = force × distance gives:

frictional force = $\dfrac{\text{work done to overcome friction}}{\text{distance moved}} = \dfrac{1163 \text{ J}}{10.2 \text{ m}} = 114 \text{N}$ [1]   *(3 marks)*

2 a When she accelerates, she loses gravitational potential energy (GPE) and gains kinetic energy. [1] When she decelerates, energy is transferred to the trampoline as she loses all her kinetic energy as she slows down to a standstill [1] and she loses gravitational potential energy as she descends to her lowest position. [1]   *(3 marks)*

b i gain of kinetic energy $\dfrac{1}{2}mv^2$ = loss of potential energy $mg\Delta h$ [1]

Rearranging this equation gives $v = \sqrt{2g\,\Delta h} = \sqrt{(2 \times 9.81 \times (0.90 - 0.16)} = 3.8 \text{m s}^{-1}$ [1]

ii Maximum energy transferred = total loss of gravitational potential energy = $mg\Delta h$ [1]

= $35 \times 9.81 \times 0.90 = 310 \text{J}$ [1] (to 2 significant figures)   *(4 marks)*

c Tension T at maximum extension = $k\Delta L = 3500 \times 0.042 = 147 \text{N}$ [1] (where extension $\Delta L = 0.042$ m)

Maximimum energy stored in each spring = $\dfrac{1}{2}T\Delta L$ = $0.5 \times 147 \times 0.042 = 3.1 \text{J}$

Total energy stored = $90 \times 3.1 \text{J} = 280 \text{J}$ to 2 significant figures [1]   *(2 marks)*

3 a Kinetic energy = $\dfrac{1}{2}mv^2 = 0.5 \times 12\,000 \times 18^2$ = 1.9 MJ [1]   *(1 mark)*

b i Distance travelled per second = 18 m therefore height gain per second = $18 \sin 5.0°$ [1]

Gain of potential energy each second = $mg\Delta h$ = $12000 \times 9.81 \times 18 \sin 5.0 = 1.9 \times 10^5 \text{J}$ [1]   *(2 marks)*

ii Rearranging $P = Fv$ gives $F = \dfrac{P}{v} = \dfrac{228\,000}{18}$ [1]

$= 1.3 \times 10^4\,\text{N}$ [1]   *(2 marks)*

iii Component of weight acting down the slope
$= mg\sin 5°$

$= 12\,000 \times 9.81 \times \sin 5 = 1.03 \times 10^4\,\text{N}$ [1]

Total resistive force = output force of engine −
component of weight acting down the slope [1]

$= 1.27 \times 10^4\,\text{N} - 1.03 \times 10^4\,\text{N} = 2.4 \times 10^3\,\text{N}$ [1]

(3 marks)

Alternative method: Power wasted due to
resistive force = output power − GPE gain
per second [1] = 228 kW − 185 kW = 43 kW.
Therefore, total resistive force

$= \dfrac{\text{power wasted}}{\text{speed}} = \dfrac{43\ \text{kW}}{18\ \text{m s}^{-1}}$ [1] $= 2.4 \times 10^3\,\text{N}$ [1]

4 a i Volume of trapped water = area × depth

$= 150 \times 10^6 \times 5.0 = 7.5 \times 10^8\,\text{m}^3$ [1]

Mass of trapped water = volume × density [1]

$= 7.5 \times 10^8 \times 1050 = 7.9 \times 10^{11}\,\text{kg}$ [1]   *(2 marks)*

ii Height change Δh of trapped water due to
release $= 0.5 \times 5.0\,\text{m} = 2.5\,\text{m}$ [1]

Loss of GPE $= mg\Delta h = 7.9 \times 10^{11} \times 9.81 \times 2.5$

$= 1.94 \times 10^{13}\,\text{J}$ [1]

Loss of GPE per second $= \dfrac{\text{loss of GPE}}{\text{time taken}}$

$= \dfrac{1.94 \times 10^{13}}{6.0 \times 3600} = 900\,\text{MW}$ [1]   *(3 marks)*

b Area of panels needed $= \dfrac{900\ \text{MW}}{220\ \text{W m}^{-1}}$

$= 4.1 \times 10^6\,\text{m}^2$ [1]   *(1 mark)*

5 a Gain of potential energy $= mg\Delta h = 1200 \times 9.81$
$\times\, 15 = 177\,\text{kJ}$ [1]

Minimum output power $= \dfrac{\text{gain of GPE}}{\text{time taken}} = \dfrac{177\ \text{kJ}}{20\ \text{s}}$

$= 8.89\,\text{kW}$ [1]

Input power $\approx \dfrac{\text{minimum output power}}{\text{efficiency}} = \dfrac{8.89\ \text{kW}}{0.58}$

$= 15\,\text{kW}$ [1]   *(3 marks)*

b Assuming the gain of kinetic energy = loss of
potential energy [1]

$\frac{1}{2}mv^2 - \frac{1}{2}mu^2 = mg\Delta h$ where $u = 1.2\,\text{m s}^{-1}$ and
$mg\Delta h = 177\,\text{kJ}$

Therefore, $0.5 \times 1200 \times v^2 - (0.5 \times 1200 \times 1.2^2)$
$= 177\,\text{kJ}$ [1]

$600v^2 = 177\,\text{kJ} + (600 \times 1.2^2) = 178\,\text{kJ}$ [1] hence

$v = \sqrt{\dfrac{178000}{600}} = 17\,\text{m s}^{-1}$ [1]   *(4 marks)*

c i Energy transferred to the water = loss of KE of
carriage in the water

$= 178\,\text{kJ} - (0.5 \times 1200 \times 11.5^2)$ [1] $= 178\,\text{kJ} -$
$79\,\text{kJ} = 99\,\text{kJ}$ [1]

ii Assume the work done to overcome resistive
forces = energy transferred to the water, then
using work done = force × distance gives

average resistive force $= \dfrac{\text{work done}}{\text{distance moved}}$

$= \dfrac{99\ \text{kJ}}{18.4\ \text{m}} = 5.4\,\text{kN}$ [2]   *(4 marks)*

6 a i Kinetic energy $= \frac{1}{2}mv^2 = 0.5 \times 250\,000 \times 150^2$
$= 2.8 \times 10^9\,\text{J}$ [1]

ii Assume work done = gain of kinetic
energy. As work done = force × distance,

force $= \dfrac{\text{gain of kinetic energy}}{\text{distance}}$ [1] $= \dfrac{2.81 \times 10^9}{2400}$

$= 1.2\,\text{MN}$ [1]   *(3 marks)*

b i Gain of GPE $= mg\Delta h = 250\,000 \times 9.81 \times 8500$
$= 2100\,\text{MJ}$ [1]

ii Kinetic energy at $250\,\text{m s}^{-1} = \frac{1}{2}mv^2 = 0.5 \times$
$250\,000 \times 250^2 = 7800\,\text{MJ}$ [1]   *(2 marks)*

c Minimum energy needed $= 2100 + 7100$
$= 9890\,\text{MJ}$ [1]

Useful energy from each kilogram of fuel
$= 0.55 \times 30\,\text{MJ} = 16.5\,\text{MJ}$ [1]

Mass of fuel used $\approx \dfrac{9900\ \text{MJ}}{16.5\ \text{MJ kg}^{-1}} = 600\,\text{kg}$ [1]

(3 marks)

7 a GPE per second transferred by the water

$= \dfrac{450\ \text{MW}}{0.80} = 563\,\text{MW}$ [1]

Therefore, $\dfrac{mg\ h}{t} = 563\,\text{MW}$ where $\Delta h = 390\,\text{m}$
and m is the mass of water that passes through
the turbine in time t. [1]

So the mass flow per second $= \dfrac{m}{t}$

$= \dfrac{\text{GPE per second transferred}}{g\Delta h}$ [1]

$= \dfrac{563 \times 10^6}{9.81 \times 390} = 1.47 \times 10^5\,\text{kg s}^{-1}$ [1]

Volume of water per second passing through the

turbines $= \dfrac{\text{mass per second}}{\text{density of water}}$ [1]

$= \dfrac{1.47 \times 10^5\ \text{kg s}^{-1}}{1000\ \text{kg m}^{-3}} = 150\,\text{m}^3\,\text{s}^{-1}$ [1]   *(6 marks)*

b i To generate 100 MJ, the GPE transferred by the
water flowing downhill $= \dfrac{100}{0.8} = 125\,\text{MJ}$. [1]

To pump the water uphill, the electrical
energy needed $= \dfrac{125\ \text{MJ}}{0.8} = 156\,\text{MJ}$. [1]

Therefore, the % of the energy supplied that
is wasted $= \dfrac{56}{156} \times 100\% = 36\%$ [1]   *(3 marks)*

ii The efficiency of generating electrical energy
is 80%. The efficiency of storing water in the
uphill reservoir must be less than 100% [1]
as energy will always be dissipated due to
friction in the bearings of the turbines and
the generators and due to resistance heating
of the electricity cables and wires. [1] As the
electrical energy generated from pumped
storage is less than 100% of 80%, the overall
efficiency of the process is less than 80%. [1]

(3 marks)

Chapter 11

1 a Density is defined as mass per unit volume of a substance [1] *(1 mark)*

b i Sphere radius $r = 7.10\,\text{mm}$

Volume V of sphere $= \frac{4}{3}\pi r^3 \times L = \frac{4}{3} \times \pi$

$\times (7.10 \times 10^{-3})^3 = 1.50 \times 10^{-6}\,\text{m}^3$ [1]

Density of metal $= \dfrac{\text{mass}}{\text{volume}} = \dfrac{11.80 \times 10^{-3}\,\text{kg}}{1.50 \times 10^{-6}\,\text{m}^3}$

[1] $= 7870\,\text{kg}\,\text{m}^{-3}$ [1] *(2 marks)*

ii % uncertainty in radius $= \pm \dfrac{0.01\,\text{mm}}{7.10\,\text{mm}} \times 100$

$= \pm 0.14\%$

Therefore, % uncertainty in volume $= 3 \times \pm 0.14\% = \pm 0.42\%$ (as $V \propto r^3$) [1]

% uncertainty in mass $= \pm \dfrac{0.04\,\text{g}}{11.80\,\text{g}} \times 100$

$= \pm 0.34\%$ [1]

Total % uncertainty $= \pm (0.42 + 0.34)$

$= \pm 0.76\%$

Uncertainty in density $= \pm \dfrac{0.76}{100} \times 7870\,\text{kg}\,\text{m}^{-3}$

$= \pm 60\,\text{kg}\,\text{m}^{-3}$ [1] *(3 marks)*

2 a Use a top pan balance to measure the mass m of the wire. Measure the length L of a sample of the wire just less than 1 m long using a metre rule. [1]

Use a micrometer to measure the diameter of the wire in 5 different places and calculate the mean diameter d. [1]

Calculate the volume V of the sample using the formula $V = \frac{1}{4}\Delta d^2 L$ then calculate the density ρ of the material using the formula $\rho = \dfrac{m}{V}$. [1] *(3 marks)*

b i Mean thickness $= 2.11\,\text{mm}$, uncertainty

$= \frac{1}{2} \times \text{range} = + 0.5 \times (2.16 - 2.06)$

$= +0.05\,\text{mm}$ [1] *(1 mark)*

ii Volume of plate $= 0.050\,\text{m} \times 0.050\,\text{m}$

$\times 2.11 \times 10^{-3}\,\text{m} = 5.3 \times 10^{-6}\,\text{m}^3$ [1]

iii Density of metal $= \dfrac{\text{mass}}{\text{volume}} = \dfrac{14.20 \cdot 10^{-3}\,\text{kg}}{5.28 \cdot 10^{-6}\,\text{m}^3}$.

[1] $= 2700\,\text{kg}\,\text{m}^{-3}$ [1] *(3 marks)*

c % uncertainty in thickness $t = \pm \dfrac{0.05}{2.11} \times 100$

$= 2.4\%$

% uncertainty in plate length $L = \pm \dfrac{0.5}{50} \times 100$

$= 1.0\%$

% uncertainty in mass $m = \pm \dfrac{0.10}{14.20} \times 100$

$= 0.7\%$ [1]

Volume $= L^2 t$ so density $\rho = \dfrac{m}{L^2 t}$

Therefore, % uncertainty in density $= (2 \times 1.0\%)$

[1] $+ 2.4\% + 0.7\% = 5.1\%$ [1]

(Note % uncertainty in L^2

$= 2 \times$ % uncertainty in L) *(3 marks)*

3 a i Spring constant $k = $ gradient of line [1]

$= \dfrac{5.00 - 0.25\,\text{N}}{490 - 300\,\text{mm}} = 25.0\,\text{N}\,\text{m}^{-1}$ [1] *(2 marks)*

ii Weight of hanger $= 0.25\,\text{N}$ therefore extension

due to hanger $= \dfrac{0.25\,\text{N}}{25.0\,\text{N}\,\text{m}^{-1}}$

$= 0.010\,\text{m} = 10\,\text{mm}$. [1] Unstretched length of spring $= 300\,\text{mm} - 10\,\text{mm} = 290\,\text{mm}$ [1] *(2 marks)*

b Extension ΔL of spring at $450\,\text{mm} = 450 - 300$

$= 150\,\text{mm}$ [1]

Energy stored in spring $= \frac{1}{2}k\Delta L^2 = 0.5 \times 25.0 \times$

0.150^2 [1] $= 0.28\,\text{J}$ [1] *(3 marks)*

4 a i Rearranging the Young modulus equation

$E = \dfrac{TL}{A\,L}$ gives $L = \dfrac{TL}{AE}$ [1] $=$

$\dfrac{250 \times 0.320}{1.23 \times 10^{-6} \times 2.1 \times 10^{11}}$ [1] $= 3.1 \times 10^{-4}\,\text{m}$ [1] *(3 marks)*

ii For tension $T = 250\,\text{N}$, work done $= \frac{1}{2}T\Delta L$

$= 0.5 \times 250 \times 3.1 \times 10^{-4} = 0.039\,\text{J}$ [1] *(1 mark)*

b i $E_K = \frac{1}{2}mv^2 = 0.5 \times 0.170 \times 1.40^2 = 0.167\,\text{J}$ [1] *(1 mark)*

ii Rebound kinetic energy $= 0.5 \times 0.170$

$\times 1.15^2 = 0.112\,\text{J}$ [1]

% kinetic energy retained $= \dfrac{0.112}{0.167} \times 100$

$= 67\%$ [1] *(2 marks)*

iii The cord was stretched beyond its limit of proportionality [1] and some of the energy supplied to the cord by the hammer was not stored as elastic energy and was dissipated to the surroundings [1] [OR the tension against extension line for unloading was below the loading line [1] and the area between the two lines corresponds to energy dissipated [1].

(2 marks)

5 a i See Figure 1. Points plotted correctly; suitable scales; best fit line. [3]

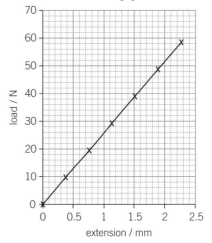

▲ **Figure 1**

ii Suitable gradient triangle shown and used correctly. [1] *(3 marks)*

Area of cross-section of wire $= \frac{1}{4}\pi d^2 = \frac{1}{4} \times \pi$

$\times (0.71 \times 10^{-3})^2 = 3.96 \times 10^{-7}\,\text{m}^2$ [1]

Gradient $= \dfrac{51.5 - 8.0\,\text{N}}{2.00 - 0.30\,\text{mm}} = 25.6\,\text{N}\,\text{mm}^{-1}$

$= 2.56 \times 10^4\,\text{N}\,\text{m}^{-1}$ [1]

$\left(\text{As the gradient} = \dfrac{T}{\Delta L}\right)$, Young modulus

equation $E = \dfrac{TL}{A\Delta L} = \text{gradient} \times \dfrac{L}{A}$

$= 2.56 \times 10^4 \times \dfrac{2.92}{3.96 \times 10^{-7}}$ [1] $= 1.9 \times 10^{11}$ Pa [1]

(5 marks)

b Tension = stress × area of cross–section so maximum load

$= 530 \times 10^6$ Pa $\times 3.96 \times 10^{-7}$ m^2 = 210 N [1]. *(1 mark)*

6 a P: The Young modulus is equal to the gradient of the initial straight section of each line [1] and P has the largest initial gradient [1]. *(2 marks)*

b i Q: A brittle material breaks without stretching significantly beyond its limit of proportionality [1] which is what happens to Q but not to P or R. [1] *(2 marks)*

ii R: A ductile material stretches considerably before it snaps without much extra force after its limit of proportionality is reached. [1] R is the most ductile because it stretches more than P or Q [1]. *(2 marks)*

c P: The energy stored per unit volume is given by the area under the line. [1] This area up to the limit of proportionality is greatest for P. [1] *(2 marks)*

7 a Tension T in each cable $= \dfrac{16300 \text{ N}}{6} = 2720$ N [1]

Area of cross-section A of each cable $= \dfrac{1}{4}\pi d^2 = \dfrac{1}{4} \times \pi \times (6.0 \times 10^{-3})^2 = 2.83 \times 10^{-5}$ m^2 [1]

Rearranging the Young modulus equation $E = \dfrac{TL}{A\Delta L}$ gives $\Delta L = \dfrac{TL}{AE}$ [1]

$= \dfrac{2720 \times 35.0}{2.83 \times 10^{-5} \times 210 \times 10^9}$ [1] $= 1.60 \times 10^{-2}$ m [1]

(5 marks)

b i Acceleration $= \dfrac{\text{change of velocity}}{\text{time taken}} = \dfrac{0.72 - 0}{8.0}$

$= 0.090$ m s^{-2} [1] *(1 mark)*

ii Total mass accelerated $= \dfrac{16300 \text{ N}}{g} = 1660$ kg

Resultant force due to extra tension
$= ma = 1660 \times 0.090 = 149$ N [1]

Extra tension in each cable $= \dfrac{149 \text{ N}}{6} = 25.0$ N

The stiffness constant k of the cable
$= \dfrac{AE}{L} = \dfrac{T}{\Delta L} = \dfrac{2720 \text{ N}}{1.60 \times 10^{-2} \text{ m}} = 1.70 \times 10^5$ N m^{-1} [1]

Therefore, the extra extension of each cable [1]
$= \dfrac{\text{extra tension}}{k} = \dfrac{25.0 \text{ N}}{1.70 \times 10^5 \text{ N m}^{-1}}$

$= 1.5 \times 10^{-4}$ m (to 2 significant figures) [1]

(4 marks)

Alternative method: The cable tension when accelerating $= mg + ma$. [1] Therefore, the increase of
$\dfrac{\text{cable tension}}{\text{cable tension at constant velocity}} = \dfrac{ma}{mg} = \dfrac{a}{g}$ [1]

The extension is proportional to the cable tension [1] therefore the increase of the extension =
$\dfrac{a}{g} \times \text{extension} = \dfrac{0.09}{9.81} \times 1.60 \times 10^{-1}$ m
$= 1.5 \times 10^{-4}$ m.] [1]

Chapter 12

1 a Resistance of wire $R = \dfrac{V}{I} = \dfrac{6.0 \text{ V}}{2.6 \text{ A}} = 2.31\,\Omega$ [1]

Area of cross-section $= \dfrac{1}{4}\pi d^2 = 0.25\pi \times (0.38 \times 10^{-3})^2 = 1.13 \times 10^{-7}$ m^2 [1]

Rearranging $\rho = \dfrac{RA}{L}$ gives $L = \dfrac{RA}{\rho}$

$= \dfrac{2.31 \times 1.13 \times 10^{-7} \text{ m}^2}{4.8 \times 10^{-7} \text{ m}}$ [1] $= 0.54$ m [1]

(4 marks)

b i $\Delta Q = I\Delta t = 2.6$ A $\times 600$ s $= 1560$ C [1]

ii Energy transferred $= IV\Delta t = 2.6$ A $\times 6.0$ V $\times 600$ s $= 9400$ J (2 s.f.) [1] *(2 marks)*

c The conduction electrons collide with positive ions in the resistor and transfer kinetic energy to them so the ions vibrate more. [1] The internal energy of the resistor increases so its temperature increases [1] and it loses energy by heating the surroundings. [1] *(3 marks)*

2 a i Resistance at 6.00 V $= \dfrac{V}{I} = \dfrac{6.00 \text{ V}}{1.20 \text{ A}} = 5.0\,\Omega$ [1]

ii Resistance at 12.00 V $= \dfrac{V}{I} = \dfrac{12.00 \text{ V}}{1.72 \text{ A}} = 6.98\,\Omega$ [1] *(2 marks)*

b i Energy transferred $= IV\Delta t = 1.72$ A $\times 12.00$ V $\times 300$ s $= 6190$ J [1]

ii $\Delta Q = I\Delta t = 1.72$ A $\times 300$ s $= 516$ C [1] *(2 marks)*

c $\dfrac{\text{Power at 12 V}}{\text{Power at 6 V}} = \dfrac{12.00 \times 1.72}{6.00 \times 1.20}$ [1] $= 2.9$ [1] *(2 marks)*

3 a See Figure 1: correctly plotted points [1] suitable scales [1] best fit line [1] *(3 marks)*

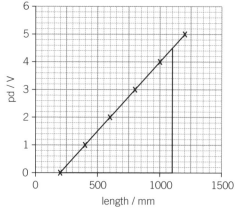

▲ **Figure 1**

b i Substituting $R = \dfrac{V}{I}$ into $\rho = \dfrac{RA}{L}$ gives $\rho = \dfrac{VA}{IL}$

Rearranging this equation gives $V = \dfrac{\rho IL}{A}$ [1] *(1 mark)*

ii Equation above is the equation of a straight line through the origin $y = mx$
where $y = V$, $x = L$, and m = gradient [1]

Gradient of graph $m = \dfrac{V}{L} = \dfrac{\rho I}{A}$ so resistivity ρ = gradient $\times \dfrac{A}{I}$ [1]

Gradient $m = \dfrac{4.50 - 0}{1100 - 200} = 5.0 \times 10^{-3}$ V mm^{-1}
$= 5.0$ V m^{-1} [1]

Area of cross-section $A = \frac{1}{4}\pi d^2 = 0.25\pi \times$
$(0.26 \times 10^{-3})^2 = 5.31 \times 10^{-8}\,m^2$ [1]

Resistivity $\rho = $ gradient $\times \frac{A}{I} = 5.0\,V\,m^{-1} \times$
$\frac{5.31 \times 10^{-8}\,m^2}{0.41\,A} = 6.5 \times 10^{-7}\,m$ [1]   (5 marks)

c   A thinner wire has a smaller cross-sectional area A
and the resistivity and the current are the same. [1]
As the graph gradient $m = \frac{\rho I}{A}$, the gradient would be
greater as A is smaller. [1]   (2 marks)

4   a   See Figure 2 for circuit diagram: correct position
of voltmeter and variable resistor [1]

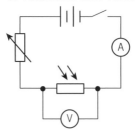

▲ Figure 2

Set up the circuit and close the switch with the
LDR in daylight. Use the variable resistor to adjust
the current to give a suitable reading on the
voltmeter without overloading the ammeter. [1]
Use the ammeter and the voltmeter to measure
the current and pd. [1] Open the switch and
cover the LDR completely. Close the switch again
and adjust the variable resistor to give the same
reading on the voltmeter as before. Measure the
current at this pd. [1] Since the pd is the same
for both sets of readings, the ratio of the dark
resistance to the daylight resistance is equal to the
daylight current /the dark current. [1]   (5 marks)

b   i   When the light intensity increases, the LDR
resistance decreases. [1] As the pd across
the LDR is constant and the LDR resistance
becomes smaller, the current in the circuit
therefore increases. [1]   (2 marks)

ii $R = \frac{V}{I} = \frac{4.5\,V}{1.2 \times 10^{-3}\,A} = 3800\,\Omega$ [1]   (1 mark)

iii   The number of charge carriers in the LDR
increases when the light intensity on the LDR
increases. [1] For any given pd across the LDR,
the current therefore becomes greater because
there are more charge carriers in the LDR. [1]
Since the resistance is equal to the pd /current,
the resistance therefore becomes smaller when
the light intensity is increased. [1]   (3 marks)

5   a   The forward voltage is the least pd across the
diode needed to make it conduct when it is in a
circuit in the forward direction. [1]   (1 mark)

b   i $\Delta Q = I\Delta t = 0.17\,A \times 10 \times 10^{-3}\,s = 1.7 \times 10^{-3}\,C$ [1]

ii $P = IV = 0.17\,A \times 2.2\,V = 0.37\,W$ [1]   (2 marks)

c   i   Maximum number of photons of energy E
emitted $= \frac{P}{E} = \frac{0.37\,W}{3.5 \times 10^{-19}} \approx 10^{18}$ [1]

ii   Some photons would be absorbed by the
LED material (or energy is dissipated by the
LED due to its resistance when current passes
through it). [1]   (2 marks)

6   a   i   The transition temperature is the temperature
at and below which the material has zero
resistivity. [1]   (1 mark)

ii   See Figure 3.

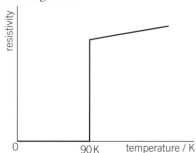

▲ Figure 3

Correct graph with vertical step at 90 K [1],
resistance below 90 K clearly zero. [1]   (2 marks)

b   i   Below the transition temperature, the resistance
R of the wire is zero so the pd across it is zero
as $V = IR$. The current is non-zero because the
circuit is a complete circuit. [1]   (1 mark)

ii   The ammeter reading decreases because
the wire has non-zero resistance and the
circuit resistance is therefore greater. [1] The
voltmeter reading is no longer zero because
the wire has non-zero resistance and a current
passes through it. [1]   (2 marks)

7   a   Resistance of wire $R = \frac{V}{I} = \frac{230\,V}{10.8\,A} = 21.3\,\Omega$ [1]

Area of cross-section $= \frac{1}{4}\pi d^2 = 0.25\pi$
$\times (0.30 \times 10^{-3})^2 = 7.07 \times 10^{-8}\,m^2$ [1]
Rearranging $\rho = \frac{RA}{L}$ gives $L = \frac{RA}{\rho}$
$= \frac{21.3 \times 7.07 \times 10^{-8}\,m^2}{5.6 \times 10^{-8}\,m}$ [1] $= 27$ m (2 s.f.) [1]   (4 marks)

b   i   Since $R = \frac{\rho L}{A}$ and the resistance and resistivity
are unchanged, then $\frac{L}{A}$ is unchanged. [1] The
diameter of X is twice that of W so the area
of cross-section of X is 4 times that of W. For
$\frac{L}{A}$ to be unchanged, the length L of X would
therefore need to be 4 times that of W [1]
which would mean the length of X would
need to be 108 m. [1]   (3 marks)

ii   As X is thicker, there could be half as many
turns per centimetre on the tube. [1] Because
X is four times longer than W, four times as
many turns round the tube would be needed
for X compared with W. [1] So the tube
would need to at least twice as long and this
is unlikely to be realistic unless the heater is
redesigned [1]. Also, the element may not be as
hot because the energy supplied to it would be
dissipated over a greater length [1]   (4 marks)

Chapter 13

1 a i Total resistance $R = \left(\dfrac{1}{4.0} + \dfrac{1}{6.0}\right)^{-1} + 3.0 = 5.4\,\Omega$ [1]

Current $= \dfrac{\text{battery emf}}{\text{total resistance}} = \dfrac{6.0\,\text{V}}{5.4} = 1.1\,\text{A}$ [1]

(2 marks)

 ii Pd across internal resistance $= 1.1\,\text{A} \times 3.0\,\Omega$
 $= 3.3\,\text{V}$ [1]

Therefore, pd across parallel combination
= battery emf − pd across internal resistance
$= 6.0 − 3.3 = 2.7\,\text{V}$ [1]   (2 marks)

 b The circuit resistance increases so the current in the battery decreases. [1] Therefore, the pd across the internal resistance decreases. [1] Since the pd across the external resistor is equal to the battery pd − the pd across the internal resistance, the pd across the battery increases. [1]   (3 marks)

2 a i pd across silicon diode in the forward direction = 0.6 V [1]

pd across resistor = battery pd − pd across diode $= 9.0\,\text{V} − 0.6\,\text{V} = 8.4\,\text{V}$ [1]   (2 marks)

 ii Current in diode = current in resistor
 $= \dfrac{\text{resistor pd}}{\text{resistance of resistor}} = \dfrac{8.4\,\text{V}}{6.8\,\text{k}} = 1.2\,\text{A}$
 (2 s.f.) [1]   (1 mark)

 b The pd across the diode is unchanged so the pd across the two parallel resistors is 8.4 V. [1] The current in each resistor is therefore 1.24 mA and therefore the current in the diode and the battery is 2.5 mA (= 2 × 1.24 mA). [1]   (2 marks)

3 a Using the circuit shown in Figure 1, the variable resistor is used to change the current which is measured using the ammeter. The battery pd is measured using the voltmeter for at least six different currents. [1] (The lamp is included to limit the maximum value of the current.)

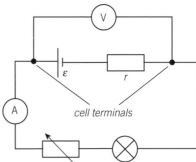

▲ Figure 1

The measurements are recorded and plotted on a graph of y = battery pd V against x = current I. The graph should be a straight line with a negative gradient in accordance with the equation $V = \varepsilon − Ir$ where ε is the battery emf and r is the internal resistance of the battery. [1] The gradient of the line is equal to $− r$ where r is the internal resistance of the battery. [1]

The internal resistance is determined by using a large section of the line to draw a gradient triangle and using the triangle to measure the change of pd ΔV and the corresponding change of

current ΔI. [1] The internal resistance r is equal to the magnitude of $\dfrac{V}{I}$. [1]   (5 marks)

 b i The external resistance R of the circuit
 $= \left(\dfrac{1}{3.0} + \dfrac{1}{9.0}\right)^{-1} = 2.25\,\Omega$. [1] Since the pd across the external resistors is 5.4 V, the circuit current $= \dfrac{5.4\,\text{V}}{2.25\,\Omega} = 2.4\,\text{A}$. [1]   (2 marks)

 ii Using the equation $\varepsilon = V + Ir$ with $\varepsilon = 6.0\,\text{V}$, $V = 5.4\,\text{V}$, and $I = 2.4\,\text{A}$ gives
 $6.0 = 5.4 + 2.4r$ [1] Therefore, $2.4r = 6.0 − 5.4$
 $= 0.60$ hence $r = \dfrac{0.60}{2.4} = 0.25\,\Omega$ [1]   (2 marks)

 iii $\dfrac{\text{Power dissipated in the internal resistance}}{\text{Power supplied by the battery}}$
 $= \dfrac{I^2 r}{I\varepsilon} = \dfrac{Ir}{\varepsilon} = \dfrac{2.4 \times 0.25}{6.0} = 0.10$ [1]   (1 mark)

4 a i Net emf = 5.0 − 2.0 = 3.0 V [1]   (1 mark)

 ii Total circuit resistance $= \dfrac{3.0\,\text{V}}{0.50\,\text{A}} = 6.0\,\Omega$ [1]

Internal resistance of B = total circuit resistance − power supply internal resistance
$= 6.0\,\Omega − 0.9\,\Omega = 5.1\,\Omega$ [1]   (2 marks)

 b i The net emf of the two batteries decreases as the mobile phone battery emf increases. [1] The circuit resistance is unchanged so the current in the circuit decreases. [1]   (2 marks)

 ii Rearranging $\Delta Q = I\Delta t$ gives $I = \dfrac{\Delta Q}{\Delta t}$
 $= \dfrac{1600\,\text{C}}{3 \times 3600\,\text{s}} = 0.15\,\text{A}$ [1]   (1 mark)

5 a Energy supplied in 1 hour $= IV\Delta t = 3.0\,\text{A} \times 6.0\,\text{V} \times 3600\,\text{s} = 65\,000\,\text{J}$ [1]   (1 mark)

 b i Power supplied by panel = 3.0 A × 6.0 V = 18.0 W

Power produced by each cell = 1.0 A × 1.2 V = 1.2 W [1]

Therefore, the number of cells $= \dfrac{18.0\,\text{W}}{1.2\,\text{W}} = 15$ [1]

Circuit diagram: the diagram should show 3 parallel rows of cells with 5 cells in each row acting in the same direction. [1]   (3 marks)

 ii The panel pd of 6.3 V when the appliance is disconnected is the net emf of the panel. [1]

When the panel is connected to the appliance, the lost voltage due to its internal resistance
$= 6.3\,\text{V} − 6.0\,\text{V} = 0.3\,\text{V}$. [1]

The current in the panel when it is connected to the appliance = 3.0 A. Therefore, the internal resistance of the panel $= \dfrac{\text{lost pd}}{\text{current}} = \dfrac{0.3\,\text{V}}{3.0\,\text{A}} = 0.1\,\Omega$ [1]   (3 marks)

6 a A potential divider is a circuit consisting of two resistors connected to a source of fixed pd. The potential difference of the source is divided or 'shared' between the two resistors in proportion to their resistances. [1] By changing the ratio of the resistances, the share of the potential difference across each resistor can be changed. [1]   (2 marks)

b The pd across the LDR = battery pd − pd across $R = 5.0 − 4.2 = 0.8$ V [1]

Therefore, the ratio of the LDR resistance to the resistance of $R = \dfrac{0.8\,\text{V}}{4.2\,\text{V}} = 0.19$ [1]

Hence, the LDR resistance = $0.19 × 1.5\,\text{k}\Omega = 290\,\Omega$ [1]   *(3 marks)*

c When the light intensity is reduced, the LDR resistance increases [1] so the pd across the LDR increases and the pd across R decreases. [1]   *(2 marks)*

d Connect the LDR to the circuit using long connecting wires. [1]

Draw radial lines at 15° intervals on a large fixed sheet of white paper from a fixed point P at the centre of the sheet and place the lamp at P. [1]

With the LDR facing the lamp directly [1], move it along one of the lines until the voltmeter reading is exactly 2.5 V and mark the position of the LDR on the line. [1]

Repeat the procedure on an adjacent line and then on successive unmarked adjacent lines until all the radial lines are marked. [1]

Since the marks all correspond to the same voltmeter reading, their position along each line indicates where the light intensity is the same along each line. [1]

Joining the marks gives a 'constant intensity' line round the lamp which is non-circular if the light intensity varies with direction. [1]

The nearer a mark on a line is to the lamp, the lower the intensity of the light emitted by the lamp in that direction. [1] (max 6)

7 a When the temperature of T is increased, its resistance decreases so the share of the battery pd across it decreases. [1] Therefore, the potential difference across the variable resistor increases. [1]   *(2 marks)*

b i The pd across T = $6.0\,\text{V} − 2.0\,\text{V} = 4.0\,\text{V}$ [1]

$$\frac{\text{resistance of the variable resistor}}{\text{resistance of T}} = \frac{\text{pd across the variable resistor}}{\text{pd across T}} = \frac{2.0}{4.0} = 0.5 \ [1]$$

So the resistance of the variable resistor = 0.5 × the resistance of T = $0.5 × 2.7\,\text{k}\Omega = 1.35\,\text{k}\Omega$ [1]   *(3 marks)*

ii If the thermistor temperature increases above −5.0°C with the variable resistor set at 1.35 kΩ, the alarm switches on because the thermistor resistance decreases so the pd across the variable resistor increases above 2.0 V. [1]

If the variable resistance is decreased with the thermistor at −5.0°C, the voltage across the variable resistor would decrease below 2.0 V [1] so the alarm would be off at −5.0°C. [1] The thermistor would need to be warmer in order for the alarm to switch on [1] (or the alarm would switch on only when the thermistor temperature is above a higher temperature than −5.0°C).   *(4 marks)*

Answers to summary questions

1.1

1 a 19n, 20p [1]

  b B (16 neutrons and 14 protons) [1]

  c $^{30}_{16}S$ [1]

2 a i 8p, 9n [1]   ii 4p, 5n [1]

  b Compared with a $^{9}_{4}Be$ nucleus, a $^{17}_{8}O$ nucleus has twice as much charge but less than twice as much mass. [1] So its specific charge $\left(= \dfrac{charge}{mass}\right)$ is greater than the specific charge of the $^{9}_{4}Be$ nucleus. [1]

3 a The charge of the nucleus is +92e. [1] Therefore, the specific charge of this nucleus $= \dfrac{92e}{238m} = \dfrac{92 \times 1.60 \times 10^{-19}\,C}{238 \times 1.67 \times 10^{-27}\,kg}$ [1] $= 3.70 \times 10^{7}\,C\,kg^{-1}$ [1]

  b The charge of the nucleus is +2e. [1] Therefore, the specific charge of this nucleus $= \dfrac{2e}{9m} = \dfrac{2 \times 1.60 \times 10^{-19}\,C}{9 \times 1.67 \times 10^{-27}\,kg}$ [1] $= 2.13 \times 10^{7}\,C\,kg^{-1}$ [1]

1.2

1 The strong nuclear force has a range of about 3–4 fm whereas the electrostatic force has an infinite range. [1]

The strong nuclear force is attractive from 2–3 fm to 0.5 fm and repulsive at < 0.5 fm whereas the electrostatic force between two protons is always repulsive. [1]

2 a A − 4 [1]   b Z − 1 [1]

3 a a = 64, b = 30, c = 0 (2 correct [1], all correct [1])

  b a = 224, b = 88, c = 2 (2 correct [1], all correct [1])

1.3

1 a Similarity: same speed in a vacuum. [1] Difference: a radio wave photon has a much longer wavelength/ shorter frequency than a light photon. [1]

  b To calculate the wavelength of a 50 keV photon, rearrange the equation $E = \dfrac{hc}{\lambda}$ to give $\lambda = \dfrac{hc}{E}$ Substituting values where $E = 50000 \times e$ $= 50000 \times 1.6 \times 10^{-19}$ J [1] gives $\lambda = \dfrac{hc}{E} = \dfrac{6.63 \times 10^{-34} \times 3.00 \times 10^{8}}{50000 \times 1.60 \times 10^{-19}} = 2.5 \times 10^{-11}$ J (answer to 2 significant figures) [1]

2 a To calculate the energy of a photon of wavelength 30 mm, we can use $E = \dfrac{hc}{\lambda} = \dfrac{6.63 \times 10^{-34} \times 3.00 \times 10^{8}}{30 \times 10^{-3}}$ $= 6.6 \times 10^{-24}$ J microwave [1]

  b i The wavelength is 10^{-8} times smaller than in **a** so the energy is 10^{8} times larger $\left(\text{because } E \propto \dfrac{1}{\lambda}\right)$. Hence, $E = 6.6 \times 10^{-24} \times 10^{8}$ J $= 6.6 \times 10^{-16}$ J [1]

    ii X-rays [1]

3 Using $E = \dfrac{hc}{\lambda}$ with $\lambda = 600$ nm gives $E = \dfrac{6.63 \times 10^{-34} \times 3.00 \times 10^{8}}{600 \times 10^{-9}} = 3.3(1) \times 10^{-19}$ J. [1] In 1 s, the LED emits a maximum of 5.0 mJ. Therefore, the maximum number of 600 nm photons that could be emitted is $\dfrac{5.0 \times 10^{-3}}{3.3(1) \times 10^{-19}\,J}$ [1] $\approx 1.5 \times 10^{16}$ [1]

1.4

1 1876 MeV (= 2 × 938 MeV) [1]

2 a Total energy = (2 × 0.511 MeV) [1] + 0.250 MeV = 1.272 MeV [1]

  b 0.636 MeV (= 1.272 MeV ÷ 2) [1]

3 a Pair production is a process in which a single photon passing near a nucleus creates a particle and its corresponding antiparticle and ceases to exist. [1]

  b Any two of: **1** A particle and its corresponding antiparticle must be created from a photon. [1] **2** The photon energy must be greater than twice the rest energy of the particle. **3** The interaction must take place near a nucleus. [1]

1.5

1 a a virtual photon [1]   b a W⁻ boson [1]

  c a W⁺ boson [1]

2 a See Topic 1.5 Figure 2a (emitted particles both correctly labelled [1], exchange particle = W⁻ [1])

  b See Topic 1.5 Figure 3b (emitted particles both correctly labelled [1])

3 a See Topic 1.5 Figure 4 (ingoing particles correctly labelled [1], outgoing particles correctly labelled [1], exchange particle = W⁺ from p to e⁻ [1])

  b A proton in the nucleus interacts with an inner-shell electron by emitting a W⁺ boson which is absorbed by the electron [1]. The proton changes into a neutron in the process [1] and the electron changes into a neutrino. [1]

2.1

1 a electron, antiproton [1]   b K⁰ meson [1]

2 a weak [1]   b strong [1]   c weak [1]

3 a A muon lasts longer than a π meson so it is less unstable. [1]

  b Similarity: they are both negatively charged. [1] Difference: a muon is less unstable than a π meson or a π⁻ meson decays into a muon. [1]

2.2

1 a i Leptons do not interact through the strong interaction whereas hadrons do. [1]

    ii A baryon decays directly or eventually into a proton whereas meson decay products do not include protons. [1]

b i meson

   ii baryon

   iii lepton (all 3 correct [2], 2 correct [1])

2 a $K^+ \rightarrow \pi^+ + \pi^\circ$ [1]

  b 219 MeV [1] (= 494 − 140 − 135 MeV) [1]

3 a Energy of each colliding proton = 1.94 GeV
(= 1.00 + 0.94 GeV) [1]

   0.12 GeV [1] ((2 × 1.94) − (4 × 0.94) GeV) [1]

  b Momentum is always conserved so the total momentum after the collision is not zero as the initial momentum is not zero. [1] The total energy is insufficient to provide the kinetic energy associated with the momentum and the rest energy of the 3 protons and the antiproton after the collision. [1]

2.3

1 Q: −1, +1, 0, −1   L: +1, −1, −1, 0 (all 6 correct [3], 5 correct [2], 4 correct [1])

2 a $\mu^- \rightarrow e^- + \nu_\mu + \bar{\nu}_e$ [2]

  b Electron lepton numbers are conserved:
0 = 1 + 0 + −1 [1]

   Muon lepton numbers are conserved:
+1 = 0 + +1 +0 [1]

   (Electron and muon lepton numbers are both conserved.)

3 a No [1] charge is not conserved in the change. [1]

  b No [1] electron lepton number is not conserved because the total is 1 before the change and −1 after the change. [1]

  c No [1] electron lepton number is not conserved because the total is +1 before the change and −1 after the change. [1]

  d No [1] electron lepton number is not conserved because the total is 0 before the change and −2 (−1 for the positron and −1 for the electron antineutrino) after the change. [1]

   Note muon lepton number is conserved because the antimuon and the antimuon neutrino both have a lepton number of −1.

2.4

1 a i $u\bar{d}$, S = 0 [1]

   ii $u d \bar{d}$, S = 0 [1]

   iii $u\bar{s}$, S = 1 [1]

  b A K$^-$ meson consists of an up antiquark and a strange quark. [1] So the antiparticle of a K$^-$ meson consists of an up quark and a strange antiquark. This combination is a K$^+$ meson. [1]

2 a See Topic 2.4 Fig 2a (u and d quarks and W$^-$ boson correctly labelled [1], β$^-$ and antineutrino correctly shown [1])

  b A down quark in a neutron emits a W$^-$ boson and changes to an up quark [1] so the neutron changes to a proton. [1] The W$^-$ quark decays into a β$^-$ particle (i.e., an electron) and an electron antineutrino. [1]

3 a S = −1 so the Σ$^-$ contains one strange quark. [1] Because the charge of the Σ$^-$ is −1 and the charge of the strange quark is $-\dfrac{1}{3}$ and there are two more quarks in the Σ$^-$ particle, [1] these two quarks must have a combined charge of $-\dfrac{2}{3}$ so they must be down quarks. Therefore, the Σ$^-$ particle must be composed of a strange quark and two down quarks. [1]

  b The total initial strangeness is zero because the initial particles are non-strange particles. [1] The strangeness of the K$^+$ is +1 by definition. The Σ$^-$ particle has strangeness −1. So the total final strangeness is zero. [1] Therefore, as the total initial strangeness is equal to the total final strangeness, strangeness is conserved. [1]

2.5

1 a The charge and the baryon number of an antiproton are both −1. Table 1 shows that both Q and B are conserved because the total of each is the same before and after the reaction. [2]

▼ **Table 1**

	Before	After
	p + p	p + p + p + p̄
Q	1 + 1	1 + 1 + 1 − 1
B	1 + 1	1 + 1 + 1 − 1

  b The total rest energy before the change is 2 × 0.94 GeV. The total after the change is 4 × 0.94 GeV. [1] Therefore, the two protons that collide must have at least 2 × 0.94 GeV of kinetic energy before the change. So each of these protons must have 0.94 GeV of kinetic energy. [1]

2 a i $-\dfrac{1}{3}$ **ii** 0

   iii +1 (all 3 correct [2], 2 correct [1])

  b i 1   **ii** −1

   iii 0 (all 3 correct [2], 2 correct [1])

3 a uds [1] Its strangeness of −1 means it contains 1 strange quark. [1] Since it is uncharged, the total charge of the other 2 quarks present must be equal and opposite to the charge of the strange quark (or −1/3). Therefore, the 2 other quarks must be an up quark and a down quark. [1]

  b Table 2 below shows the numbers for charge Q, baryon number B, and strangeness S. The results show that X is uncharged and is not a baryon. [1] It does have non-zero strangeness so it cannot be a lepton. [1] Therefore, it must be a strange uncharged meson. [1] Because its strangeness is +1 and it is uncharged, it must consist of a strange antiquark and a down quark. So it is a K^0 meson [1]

▼ **Table 2**

	Before		After
	π$^-$ + p	→	Λ^0 + X
Q	−1 + 1		0 + 0
B	0 + 1		1 + 0
S	0 + 0		−1 + 1

3.1

1　The work function is the minimum energy an electron must have to escape from the surface of the metal. [1]

2　When a conduction electron at or near the surface of a metal absorbs a photon from the incident light, it gains energy equal to the photon energy hf. [1] The conduction electron can only escape from the metal if its energy is greater than or equal to the work function ϕ of the metal. [1] Therefore, in order for the electron to escape, $hf \geq \phi$. Therefore, the frequency of the incident radiation must be greater than or equal to $\frac{\phi}{h}$. [1]

3　a　The work function of the metal is equal to the energy of a photon of wavelength 390 nm. Therefore, the work function is equal to 5.1×10^{-19} J. [1]

　　b　A photon of wavelength more than 390 nm will have a frequency and therefore an energy less than a photon of wavelength 390 nm. [1] Therefore, a photon of wavelength less than 390 nm will not give a conduction electron enough energy to overcome the work function of the metal and escape. [1]

3.2

1　The stopping potential of a metal is the minimum potential needed to stop photoelectric emission from the metal. [1]

2　a　i　Photoelectric emission of electrons takes place from the cathode because each emitted electron absorbs a single photon and gains enough energy to leave the cathode. [1] The emitted electrons reach the anode and flow through the ammeter round the circuit. The microammeter current is due to this flow of electrons round the circuit. [1]

　　　ii　The flow of charge Q in 1 second = current × time = $0.85\,\mu\text{A} \times 1$ second = $0.85\,\mu\text{C}$ [1]

　　　　The number of electrons per second
　　　　$= \dfrac{\text{charge flow } \overline{Q}}{e} = \dfrac{0.85 \times 10^{-6}}{1.60 \times 10^{-19}} = 5.3 \times 10^{12}\,\text{s}^{-1}$ [1]

　　b　The current through the microammeter is proportional to the intensity of the incident light. [1] This is because the number of electrons from the cathode is proportional to the number of photons incident on the cathode. [1] As the light intensity is reduced to zero the current decreases to zero in proportion to the light intensity. [1]

3　a　Photon energy $= \dfrac{hc}{\lambda} = \dfrac{6.63 \times 10^{-34} \times 3.00 \times 10^{8}}{430 \times 10^{-9}}$
　　　$= 4.63 \times 10^{-19}\,\text{J} = \dfrac{4.63 \times 10^{-19}}{1.60 \times 10^{-19}} = 2.89\,\text{eV}$ [1]

　　b　Maximum kinetic energy of a photoelectron
　　　$= hf - \phi = 2.89 - 1.30\,\text{eV}$ [1] $= 1.59\,\text{eV}$ [1]

3.3

1　Ionisation is the process of adding or removing an electron from an uncharged atom so the atom becomes charged. [1] Excitation is a process in which

an electron in an atom gains sufficient energy to move away from the nucleus to an electron shell further away from the nucleus. [1]

2　a　2.18×10^{-18} J [1] (= $13.6 \times 1.60 \times 10^{-19}$ J)

　　b　1.6 eV [1] (= 16.2 − 1.0 − 13.6 eV) [1]

3　a　The external electron collides with an electron in one of the electron shells of the atom and gives the atomic electron enough kinetic energy [1] to move to an electron shell further away from the nucleus. [1]

　　b　The slow-moving electron does not have enough kinetic energy to enable any of the atomic electrons to move to an outer shell. So it cannot cause excitation of the atom. [1]

3.4

1　a　≈ 7.2 MeV [1]

　　b　i　10. [1] 4 from the 7.2 MeV level to each lower level, 3 from the next level down to each lower level, 2 from the next level below to each lower level, and 1 from the 1st excited level to the ground state. (all correct [2], all except one correct [1])

　　　ii　The photon emitted when the atom de-excites to the ground state from the 7.2 MeV energy level. [1]

3.5

1　Line emission spectrum is a spectrum consisting of discrete lines of different wavelengths. It is produced when gas atoms become excited and then de-excite and emit photons. [1] The photons have specific energies because an excited gas atom has different energy levels. [1] Each photon is released when an electron in the excited atom moves to a lower energy level in the atom. [1]

2　a　Photon energy $= \dfrac{hc}{\lambda} = \dfrac{6.63 \times 10^{-34} \times 3.00 \times 10^{8}}{656 \times 10^{-9}}$
　　　$= 3.03 \times 10^{-19}$ J [1]

　　b　The energy of a 486 nm photon
　　　$= \dfrac{hc}{\lambda} = \dfrac{6.63 \times 10^{-34} \times 3.00 \times 10^{8}}{486 \times 10^{-9}} = 4.09 \times 10^{-19}$ J [1]

　　　Therefore, the energy difference ΔE between X and Y $= 4.09 \times 10^{-19}$ J $- 3.03 \times 10^{-19}$ J $= 1.06 \times 10^{-19}$ J [1]

　　　The wavelength of a photon released in an electron transition from X to Y
　　　$= \dfrac{h}{\Delta E} = \dfrac{6.63 \times 10^{-34} \times 3.00 \times 10^{8}}{1.06 \times 10^{-9}} = 1.88 \times 10^{-6}$ m [1]

3.6

1　Matter particles have a wave-like nature as well as a particle-like nature. [1] Their wave-like behaviour is characterised by their de Broglie wavelength which is related to their momentum, p, by means of the equation $\lambda = \dfrac{h}{p}$. [1]

2　a　de Broglie wavelength $\lambda = \dfrac{h}{mv}$
　　　$= \dfrac{6.63 \times 10^{-34}}{9.11 \times 10^{-31} \times 2.0 \times 10^{7}}$ [1] $= 3.64 \times 10^{-11}$ m [1]

b speed $v = \dfrac{h}{m\lambda} = \dfrac{6.63 \times 10^{-34}}{1.67 \times 10^{-27} \times 3.64 \times 10^{-11}}$ [1]

$= 1.09 \times 10^4 \text{ ms}^{-1}$ [1]

4.1

1 a electromagnetic; transverse

b electromagnetic; transverse

c mechanical; longitudinal

d mechanical; longitudinal (all correct [3], 3 correct [2], 2 correct [1])

2 a A longitudinal wave is a wave in which the vibrations of the particles are parallel to the direction in which the wave travels [1]

b Each time the end of the coil moved forwards, the coils at that end are compressed and they push on the nearby coils further along the slinky. As a result, the nearby coils push on the coils further along so the compression moves along the slinky. [1] When the end of the slinky is pulled backwards, the coils at the end move backwards and create a rarefaction, pulling the nearby coils back which pulls the coils back further away. [1] As a result, the earlier compression is followed by a rarefaction. Repeating the process continually sends a series of compressions and rarefactions along the slinky. [1]

3 a 1 The vibrations of a transverse wave are perpendicular to the direction in which the wave travels whereas the vibrations of a longitudinal wave are parallel to the direction in which the wave travels.

2 Transverse waves can be polarised whereas longitudinal waves cannot. [2]

b i Transverse wave shown and direction correct. [1]

ii X on a crest and direction correctly shown. [1]

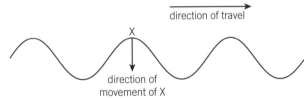

▲ Figure 1

4.2

1 a $f = \dfrac{c}{\lambda} = \dfrac{3.00 \times 10^8}{320 \times 10^{-9}} = 9.4 \times 10^{14}\text{ Hz}$ [1]

b $\lambda = \dfrac{c}{f} = \dfrac{1500}{20 \times 10^3} = 0.075\text{ m}$ [1]

2 a The frequency of a progressive wave is the number of complete waves passing a point per second. [1]

b Period $= \dfrac{1}{f} = \dfrac{1}{1.6} = 0.63\text{ s}$ (to 2 significant figures) [1]

$\lambda = \dfrac{c}{f} = \dfrac{2.9}{1.6} = 1.8\text{ m}$ [1]

3 a i $30°\left(= \dfrac{360°}{12}\right.$ as there are 12 intervals from O to S which are in phase with each other$\left.\right)$

ii $\dfrac{\pi}{6}$ (or 0.52) rad [1]

b π rad $\left(= 6\text{ intervals} \times \dfrac{\pi}{6}\right)$ [1]

c 1 They have the same amplitude and this is unchanged half a cycle later. [1]

2 Their phase difference is π rad $\left(= 6\text{ intervals} \times \dfrac{\pi}{6}\right)$ and this is constant so is unchanged. [1]

3 P is at negative maximum displacement and R is at positive maximum displacement. Half a cycle later, P is at positive maximum displacement and R is at negative maximum displacement. [1]

4.3

1 a See Figure 1 [1]

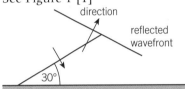

▲ Figure 1

2 a/b See Figure 2 ([1] for each part correct)

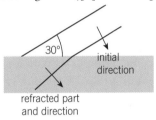

▲ Figure 2

4.4

1 a The principle of superposition states that when two waves meet, the total displacement at a point is equal to the sum of the individual displacements at that point. [1]

b The crests and troughs of a progressive wave travel along the rope whereas the crests and troughs of a stationary wave do not move along the rope. [1]

2 If the frequency is increased, the wavelength of the waves decreases. [1] The points of cancellation and reinforcement therefore become closer together. [1]

3 a Coherent sources emit waves of the same frequency with a constant phase difference. [1]

b The receiver signal is a maximum at points along the line where the microwaves from one slit arrive in phase with the microwaves from the other slit so they reinforce each other. [1] These points are regularly spaced along XY. [1] The receiver signal is a minimum midway between the maxima because at the minima, the microwaves from each slit are out of phase by 180° so they cancel each other out. [1]

4.5

1 a Two progressive waves form a stationary wave pattern when they pass through each other if they have the same frequency and they travel in opposite directions through each other. [1]

b i $\lambda = 2L = 0.60 \times 2 = 1.20\,\text{m}$ (because the string has only 2 nodes which are at its ends. Therefore, the string length $L = \dfrac{\lambda}{2}$ where λ is the wavelength.) [1]

ii $v = f\lambda = 150 \times 1.20 = 180\,\text{m s}^{-1}$ [1]

2 a The amplitude of a progressive wave is the same at all positions along the wave. [1] The amplitude of a stationary wave varies with position and is greatest at the antinodes and zero at the nodes. [1]

b The phase difference between two different points along a progressive wave increases with the distance between the two points from zero to 2π for each cycle of the wave. [1]

For a stationary wave, the phase difference is π for any two points separated by an odd number of nodes and zero for any two points separated by an even number of nodes. [1]

3 Sound waves travelling along the pipe reflect at the closed end and the reflected waves pass through the sound waves from the loudspeaker. [1] The two sets of waves have the same frequency and therefore form a stationary wave pattern in the pipe which makes the pipe resonate with sound. [1]

This happens at specific frequencies which produce a node at the closed end of the pipe and an antinode just outside the open end. [1]

4.6

1 a $\lambda = 2L = 2 \times 0.76\,\text{m} = 1.52\,\text{m}$ [1]

b Rearranging $f_1 = \dfrac{1}{2L}\sqrt{\dfrac{T}{\mu}}$ gives

$\mu = \dfrac{T}{(2Lf_1)^2} = \dfrac{64}{(2 \times 0.760 \times 280)^2}$ [1]

$= 3.5(3) \times 10^{-4}\,\text{kg m}^{-1}$ [1]

2 As T and μ are unchanged the equation $f_1 = \dfrac{1}{2L}\sqrt{\dfrac{T}{\mu}}$ means that $f \times L = $ constant. [1]

Therefore, the length L for a frequency of $384\,\text{Hz}$ is given by the equation $384\,L = 280 \times 0.760$ [1]

Hence, $L = \dfrac{280 \times 0.760}{384} = 0.554\,\text{m}$ [1]

3 Mass per unit length $\mu = $ density $\rho \times$ area of cross section $\left(\dfrac{\pi d^2}{4}\right)$ where d is the diameter.

Therefore, $\dfrac{\mu_x}{\mu_y} = \dfrac{\dfrac{\rho_x \pi d_x^2}{4}}{\dfrac{\rho_y \pi d_y^2}{4}} = \dfrac{8\rho_y}{\rho_y} \times \dfrac{\dfrac{\pi(0.5d_y^2)}{4}}{\dfrac{\pi d_y^2}{4}} = 8 \times (0.5)^2$

$= 2.0$ [1]

Because tension T and length L are constant, the equation

$f_1 = \dfrac{1}{2L}\sqrt{\dfrac{T}{\mu}}$ means that $f \times \mu^{\frac{1}{2}} = $ constant [1]

Therefore, $f_y = f_x \times \sqrt{\dfrac{\mu_x}{\mu_y}} = 100 \times \sqrt{2.0} = 141\,\text{Hz}$ [1]

4.7

1 $0.52\,\text{V}\ (= 0.5 \times 5.2\,\text{cm} \times 0.2\,\text{V cm}^{-1})$ [1]

2 Period $T = \dfrac{5.0\,\text{ms cm}^{-1} \times 6.4\,\text{cm}}{4} = 8.0\,\text{ms}$ [1]

Therefore, $f = \dfrac{1}{T} = \dfrac{1}{8.0\,\text{ms}} = 130\,\text{N (2 s.f)}$ [1]

5.1/5.2

1 $n_1 \sin\theta_1 = n_2 \sin\theta_2$ gives $1 \sin 30.0 = 1.52 \sin\theta_2$. [1]

Therefore, $\sin\theta_2 = \dfrac{\sin 30.0}{1.52}$ so $\theta_2 = 19.2°$ [1]

2 $n_1 \sin\theta_1 = n_2 \sin\theta_2$ gives $1.52 \sin\theta_1 = 1 \sin 70.0$ [1]

Therefore, $\sin\theta_1 = \dfrac{\sin 70.0}{1.52}$ so $\theta_1 = 38.2°$ [1]

3 a The light ray was not refracted because it entered the block along the normal. [1]

b At C, $\theta_1 = 35°$. Therefore, $n_1 \sin\theta_1 = n_2 \sin\theta_2$ gives $1.54 \sin 35 = 1 \sin\theta_2$. Therefore, $\sin\theta_2 = 1.54 \sin 35$ [1] so $\theta_2 = 62°$. [1] So the deflection of the light ray $= 62 - 35 = 27°$. [1]

4 $n_1 \sin\theta_1 = n_2 \sin\theta_2$ gives $1.33 \sin 40.0 = 1.50 \sin\theta_2$. [1]

Therefore, $\sin\theta_2 = \dfrac{1.33 \sin 40.0}{1.50} = 0.570$ so $\theta_1 = 34.7°$. [1]

5.3

1 a $n_1 \sin\theta_1 = n_2 \sin\theta_2$ gives $1.54 \sin c = 1 \sin 90$. [1]

Therefore, $\sin c = \dfrac{1}{1.54} = 0.649$ so $c = 40.5°$ [1]

b $n_1 \sin\theta_1 = n_2 \sin\theta_2$ gives $1.54 \sin c = 1.33 \sin 90.0$ [1]

Therefore, $\sin c = \dfrac{1.33}{1.54} = 0.864$ so $c = 59.7°$ [1]

2 a The cladding is designed such that where two fibres are in contact, their cores are not in contact. If the cores were in contact, light would pass between the two fibres, causing signals to be insecure and also light loss. [1] The cladding needs to have a lower refractive index than the core so that total internal reflection of light rays in the core can occur at the boundary. [1] The cores need to be narrow so the difference in the distance travelled by the axial rays and non-axial rays is as little as possible otherwise light pulses would become longer as they travel along the fibre. [1]

b $n_1 \sin\theta_1 = n_2 \sin\theta_2$ gives $1.51 \sin c = 1.42 \sin 90.0$ [1]

Therefore, $\sin c = \dfrac{1.42}{1.51} = 0.940$ so $c = 70.1°$ [1]

3 a A coherent bundle of fibres is a bundle in which the fibre ends are in the same relative positions at each end of the bundle. [1]

b Two fibre bundles are needed because one bundle is used to send light into the body cavity to illuminate it and the other bundle, the coherent bundle, is used to observe the cavity. [1] A lens over the end of the coherent bundle forms an image of the body cavity on the end of the coherent bundle. The image is observed at the other end of the coherent bundle. [1]

5.4

1 a The fringe spacing increases so the fringes become further apart. [1]

b The fringes would be blue and they would be closer together. [1]

2 Fringe spacing $= \dfrac{4.5\,\text{mm}}{4} = 1.125\,\text{mm}$ [1]

$\lambda = \dfrac{ws}{D} = \dfrac{1.125\times10^{-3}\times0.40\times10^{-3}}{0.85}$ [1] $= 5.30\times10^{-7}\,\text{m}$

$= 530\,\text{nm}$ (2 s.f.) [1]

3 The fringes would be a different colour because the wavelength is different. [1] The fringes would be further apart because the fringe spacing is larger for longer wavelengths. [1]

5.5

1 a Two or more light sources are said to be coherent sources if they emit light waves of the same frequency with a constant phase difference. [1] Note: A laser is a coherent light source because it emits light waves in phase with one another.

b A lamp filament emits light waves at random so the light waves from two filament lamps can never have a constant phase difference. [1] Therefore, two filament lamps can never produce a double slit interference pattern, as the points of cancellation and reinforcement would move about at random. [1]

2 $w = \dfrac{\lambda D}{s} = \dfrac{635\times10^{-9}\times1.90}{0.45\times10^{-3}}$ [1] $= 2.7\times10^{-3}\,\text{m}$ [1]

3 The fringes all become brighter and the colour of the central fringe changes from red to white. [1] The fringe separation decreases [1] and each non-central fringe becomes white in the middle and tinged with blue on the side nearest the central fringe and red on the other side. [1]

5.6

1 a The pattern consists of alternate bright and dark fringes. [1] The central bright fringe is brighter and twice as wide as each of the other bright fringes [1] which are less and less bright the further they are from the central fringe. [1]

b See Topic 5.6 Figure 1. (Central max is at least 3 times higher than the 1st max either side [1], central fringe is twice the width of the other bright fringes [1])

2 a The bright fringes become less intense because less light passes through the narrower slit and the diffracted light spreads out more as they diffract more. [1] The increase in diffraction also causes the bright fringes to become wider and further apart. [1]

b The colour of the fringes changes from red to blue and the amount of diffraction decreases as the wavelength of blue light is less than that of red light. [1] As a result of the decrease in diffraction, the bright fringes become closer together and narrower. [1]

5.7

1 a The grating has 600 000 lines per metre. Hence,

$d = \dfrac{1}{600\,000}\,\text{m} = 1.67\times10^{-6}\,\text{m}$ [1]

Since $\lambda = d\sin\theta$ for the 1st order beam, $\sin\theta = \dfrac{\lambda}{d}$

$= \dfrac{630\times10^{-9}}{1.67\times10^{-6}} = 0.377$ [1] which gives $\theta = 22°$ (2 s.f.) [1]

b maximum order number $= \dfrac{1.67\times10^{-6}}{630\times10^{-9}} = 2.65$ [1]

which rounded down $= 2$ [1]

2 $\lambda = d\sin 28°$ for the 1st order beam so the maximum order number $= \dfrac{d}{\lambda}$ rounded down ($= 1 \div \sin 28°$) $= 2.13$ rounded down $= 2$ [1]

The angle of diffraction of the 2nd order beam is given by the equation $2\lambda = d\sin\theta_2$ hence $\sin\theta_2 = 2\lambda/d$ $= 2/2.13 = 0.9389$ [1] which gives $\theta_2 = 70°$ [1]

6.1

1 a See Figure 1. [1]

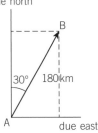

▲ **Figure 1**

b i Component due East $= 180\sin 30 = 90\,\text{km}$

ii Component due North $= 180\cos 30 = 156\,\text{km}$ [1]

2 a Resultant $= 11.0 - 5.0 = 6.0\,\text{N}$ in the direction of the 11 N force [1]

b See Figure 2 for a sketch diagram.

Magnitude of the resultant $= \sqrt{F_1^2 + F_2^2}$ $= \sqrt{5.0^2 + 11.0^2} = 12.1\,\text{N}$ [1]

Direction of resultant is given by $\tan\theta = \dfrac{F_2}{F_1}$ which gives $\tan\theta = \dfrac{11.0}{5.0} = 2.20$ hence $\theta = 66°$ (2 s.f.) [1] between F_1 and the resultant.

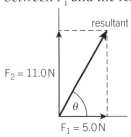

▲ **Figure 2**

3 See Figure 3 for a sketch diagram.

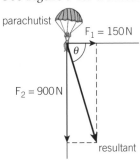

▲ **Figure 3**

Magnitude of the resultant
$= \sqrt{F_1^2 + F_2^2} = \sqrt{150^2 + 900^2} = 912\,N$ [1]

Direction of resultant is given by $\tan\theta = \dfrac{F_2}{F_1}$ [1] which gives $\tan\theta = \dfrac{900}{150} = 6.00$ hence $\theta = 80.5°$
between F_1 (which is horizontal) and the resultant. [1]

6.2

1 **a** Magnitude of resultant $= 6.6\,N = \sqrt{3.2^2 + 5.8^2}\,N$ [1]

   Angle of resultant below the horizontal
   $= 29°$ [1] $\left(= \tan^{-1}\dfrac{3.2}{5.8}\right)$ [1]

   b The third force must be equal and opposite to the resultant in **a** and therefore the third force must be 6.6 N in a direction of 29° above the horizontal (i.e. in the opposite direction to the resultant in part **a**). [1]

2 Resolving W parallel and perpendicular to the slope correctly [1] gives:

   a horizontally: $F = W\sin\theta = 8.6\sin 40 = 5.5\,N$ [1]

   b vertically: $S = W\cos\theta = 8.6\cos 40 = 6.6\,N$ [1]

Note: The lines of action of the three forces in Figure 3 have no overall turning effect because they act through the same point.

3 See Figure 1.

   Resolving T_1 and T_2 vertically and horizontally gives:

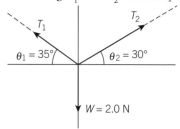

$\theta_1 = 35°$ $\theta_2 = 30°$

$W = 2.0\,N$

   Horizontally: $T_1\cos 35 = T_2\cos 30$ therefore
   $T_1 = \dfrac{T_2\cos 30}{\cos 35} = 1.057\,T_2$ [1]

   Vertically: $T_1\sin 35 + T_2\sin 30 = W$ therefore
   $1.057\sin 35\,T_2 + 0.500\,T_2 = W$ [1]

   Therefore, $0.606\,T_2 + 0.500\,T_2 = W$ which gives
   $1.106\,T_2 = 2.0$ so $T_2 = \dfrac{2.00}{1.106} = 1.8(1)\,N$ [1] and
   $T_1 = 1.057\,T_2 = 1.9(1)\,N$ [1]

6.3

1 **a** See Figure 1 (figure not to scale). [1]

0cm 10cm   25cm   50cm   90cm

2.0N   3.0N   W

▲ **Figure 1**

   b Taking moments about the pivot:

   Sum of the clockwise moments
   $= W \times (0.90 - 0.50) = 0.40\,W$ [1]

   Sum of the anticlockwise moments

   $= [2.0 \times (0.50 - 0.10)] + [3.0 \times (0.50 - 0.25)]$
   $= 0.80 + 0.75$ [1]

Applying the principle of moments gives
$0.40\,W = 0.80 + 0.75 = 1.55$

Hence, $W = \dfrac{1.55}{0.40} = 3.9\,N$ (to 2 significant figures) [1]

2 See Figure 2 for a sketch of the arrangement. Assume the force F of the adult on the seesaw is vertically downwards.

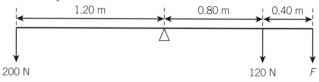

1.20 m   0.80 m   0.40 m

200 N   120 N   F

▲ **Figure 2**

Taking moments about the pivot,

Sum of the clockwise moments $= (120 \times 0.80) + 1.20\,F = 96.0 + 1.20\,F$ [1]

Sum of the anticlockwise moments $= (200 \times 1.20)$
$= 240\,Nm$ [1]

Applying the principle of moments gives
$96.0 + 1.20\,F = 240$

Therefore, $1.20\,F = 240 - 96 = 144$ which gives
$F = \dfrac{144}{1.20} = 120\,N$ (to 2 significant figures) [1]

3 See Figure 3 where F is the force to be calculated [1]

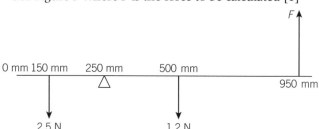

0 mm 150 mm   250 mm   500 mm

F

950 mm

2.5 N   1.2 N

Taking moments about the pivot,

Sum of the clockwise moments
$= 1.2 \times (0.500 - 0.250) = 0.300\,Nm$ [1]

Sum of the anticlockwise moments
$= [2.5 \times (0.250 - 0.150)] + [F \times (0.950 - 0.250)]$
$= 0.250 + 0.70\,F$ [1]

Applying the principle of moments gives
$0.70F + 0.250 = 0.300$

Hence, $0.70\,F = 0.300 - 0.250 = 0.050$ therefore
$F = \dfrac{0.050}{0.70} = 0.071\,N$ (to 2 significant figures) [1]

6.4/6.5

1 See Figure 1 [1]

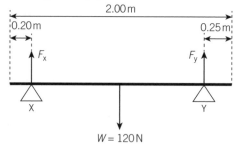

2.00 m

0.20 m   0.25 m

F_x F_y

X   Y

$W = 120\,N$

▲ **Figure 1**

Taking moments about Y to find the force F_x at X,

Sum of the clockwise moments $= F_x \times 1.55$ since XY
$= (2.00 - 0.25 - 0.20)$ [1]

Sum of the anticlockwise moments
= $120 \times (1.00 - 0.25) = 90\,\text{Nm}$ [1]

Applying the principle of moments gives $1.55 F_x = 90$

Hence, $F_x = \dfrac{90}{1.55} = 58\,\text{N}$ (to 2 significant figures) [1]

To find the force F_y at Y, since $F_x + F_y = W$, then

$F_y = W - F_x = 120 - 58 = 62\,\text{N}$ [1]

2 Taking moments about Q to find the force F_p at P, since the perpendicular distances from Q of the forces are 11.0 m for the force at P, 5.5 m for the weight of the span (6400 N) and 7.0 m (= 11.0 − 4.0 m) for the load weight (500 N) then:

Sum of the clockwise moments = $F_p \times 11.0$ [1]

Sum of the anticlockwise moments
= $(500 \times (11.0 - 4.0) + (6400 \times 5.5) = 38\,700\,\text{Nm}$

Applying the principle of moments gives
$11.0 F_p = 38700$ [1]

Hence, $F_p = \dfrac{38700}{11.0} = 3520\,\text{N}$ (to 3 significant figures) [1]

The total weight on the pillars = 6400 + 500 = 6900 N

To find the force F_q at Q, since $F_p + F_q = 6900$, then $F_q = W - F_p = 6900 - 3520 = 3380\,\text{N}$ [1]

3 a See Figure 2. The force on the board at A acts vertically downward to stop the board turning about P and tipping into the pool. [1]

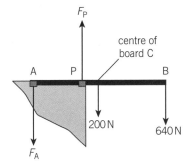

▲ **Figure 2**

b i The perpendicular distance of the force from P is PB = 1.50.

AP = AB − PB = 2.50 − 1.50 = 1.00 m
PC = AC − AP = 1.25 − 1.00 = 0.25 m

Taking moments about P to find the force at A, F_A, gives:

Sum of the clockwise moments = (200 N × 0.25 m) + (640 N × 1.50 m) = 1010 Nm [1]

Sum of the anticlockwise moments = $F_A \times 1.00$ m [1]

Applying the principle of moments gives
$F_A = \dfrac{1010\,\text{N m}}{1.00\,\text{m}} = 1010\,\text{N}$ downwards. [1]

ii To find the force F_p at P use
$F_p = F_A + 200\,\text{N} + 640\,\text{N} = 1010 + 840 = 1850\,\text{N}$ [1]

4 A sidewind exerts a horizontal force on the side of the lorry. The line of action of this force is above the ground and so the force has a turning effect about the points where the wheels on the other side of the lorry are in contact with the road. [1] The turning

effect makes the lorry unstable as it could make the lorry topple over. [1]

5 Pulling a drawer out shifts the centre of mass of the cabinet in the direction in which the drawer is pulled out. [1] Pulling a second drawer out would shift the centre of mass even further and may even cause the line of action of the weight of the filing cabinet to act beyond the base of the cabinet. [1] If this happens, the moment of the weight about the edge of the filing cabinet would be the only moment acting on the filing cabinet so the cabinet would topple over. [1]

6 The maximum angle of the slope for no toppling is such that the centre of mass is directly above the lower wheel, as shown in Figure 3. [1] At this position, the angle of inclination θ of the slope is given by $\tan\theta = \dfrac{0.5 \times 1.7}{0.9} = 0.944$ [1] which gives $\theta = 43°$. [1]

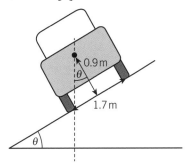

▲ **Figure 3**

6.6/6.7

1 a See Figure 1. M is the midpoint of the plank. [1]

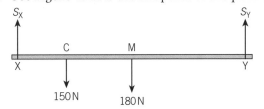

▲ **Figure 1**

b XY = 4.00 − 0.20 = 3.80 m, XC = 0.90 − 0.10 = 0.80 m

CY = 3.80 − 0.80 = 3.00 m, MY = 2.0 − 0.10 = 1.90 m

Taking moments about Y gives:

Clockwise moments = $(S_X \times XY) = 3.80 S_X$ [1]

Anticlockwise moments = (150 N × CY) + (180 N × MY) [1]

= (150 × 3.00) + (180 × 1.90) = 792 Nm

Since the plank is in equilibrium:

the sum of the clockwise moments = the sum of the anticlockwise moments

Therefore, $3.80 S_X = 792$ hence $S_X = \dfrac{792}{3.80} = 208\,\text{N}$ [1]

Considering all the force components are vertical therefore

$S_X + S_Y = 150 + 180 = 330\,\text{N}$
So, $S_Y = 330 - 208 = 122\,\text{N}$ [1]

2 a See Figure 2a: F = force due to the wall, S = support force due to floor, W = weight [1]

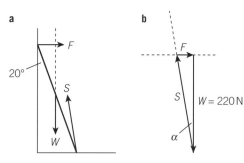

▲ Figure 2

b The distance from the top of the ladder to the floor = $L\cos 20$ where L is the length of the ladder. The distance from the bottom of the ladder to the wall = $L\sin 20$.

Taking moments about the bottom of the ladder (length L) gives:

$FL\cos 20$ [1] $= \frac{1}{2}WL\sin 20$ [1] therefore

$$F = \frac{\frac{1}{2}W\sin 20}{\cos 20} = 40\,\text{N}$$

Since F and W are perpendicular, $S^2 = F^2 + W^2 = 220^2 + 40^2$ which gives $S = 224\,\text{N}$. [1]

Resolving S vertically and horizontally gives $S\cos\alpha = W$ and $S\sin\alpha = F$.

Therefore, $\tan\alpha = \frac{F}{W} = 10.3°$ [1]

Alternative method for part b: S acts through the point where lines of action of W and F intersect on a scale diagram of the ladder at 20° to the wall. The direction of S can therefore be measured from the scale diagram and then used to draw the triangle of forces as in Figure 2b. Using Figure 2b gives $\alpha = 10.5°$ and $S = 224\,\text{N}$.

3 a See Figure 3. [1]

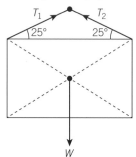

▲ Figure 3

b The tension in each section of the cord is the same because their horizontal components are equal and opposite to each other and the angle between each cord and the horizontal is the same (i.e., $T_1\cos 25 = T_2\cos 25$). [1]

The sum of their vertical components is equal and opposite to the weight (i.e., $T_1\sin 25 + T_2\sin 25 = W$). [1]

Hence, $2T_1\sin 25 = W$ so $T_1 = T_2$

$= \frac{W}{2\sin 25} = \frac{21}{2\sin 25} = 25\,\text{N}$ [1]

7.1

1 a $28\,\text{m s}^{-1}\left(= \frac{42\,000\,\text{m}}{1500\,\text{s}}\right)$ [1]

b Time taken for 2nd part of journey
$$= \frac{\text{distance}}{\text{speed}} = \frac{20\,000\,\text{m}}{20\,\text{ms}^{-1}} = 1000\,\text{s}$$ [1]

Average speed $= \frac{\text{total distance}}{\text{total time}} = \frac{62\,000\,\text{m}}{2500\,\text{s}}$ [1]
$= 25\,\text{m s}^{-1}$ (to 2 significant figures) [1]

2 a Speed $= \frac{2\pi r}{t} = \frac{2\pi \times 6.36 \times 10^6\,\text{m}}{24 \times 3600\,\text{s}}$ [1] $= 460\,\text{m s}^{-1}$
(to 2 significant figures) [1]

b i The satellite next passes over the equator in the same direction 2 hours later when it has moved through 360°. [1]

ii In 2 hours, the Earth has rotated through $\frac{1}{12}$th of a rotation hence [1]

$PQ = \frac{1}{12}$th of the circumference $= \frac{1}{12} \times 2\pi r$

$= \pi \times \frac{6360}{6}$ [1] $= 3300\,\text{km}$ (to 2 significant figures) [1]

3 a The speed of an object is change of distance per unit time. Velocity is change of displacement per unit time. [1] Displacement is distance in a certain direction. Therefore, velocity is speed in a certain direction. [1]

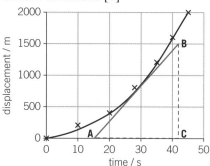

▲ Figure 1

b i See Figure 1: [1] suitable scales [1] correctly plotted points [1] best-fit curve.

ii Using the gradient triangle ABC, time taken
$= AC = 42 - 15 = 27\,\text{s}$,

distance moved $= BC = 1500 - 0 = 1500\,\text{m}$. [1]

Therefore, speed $= \frac{\text{distance}}{\text{time}} = \frac{1500\,\text{m}}{27\,\text{s}} = 56\,\text{m s}^{-1}$. [1]

7.2

1 $u = 26\,\text{m s}^{-1}$, $v = 0$, $t = 6.7\,\text{s}$
$$a = \frac{v - u}{t} = \frac{0 - 26}{6.7}$$ [1] $= -3.9\,\text{m s}^{-2}$ [1]

2 $u = 0$, $v = 1.8 \times 10^7\,\text{m s}^{-1}$, $t = 51\,\text{ns}$
$$a = \frac{v - u}{t} = \frac{1.8 \times 10^7 - 0}{51 \times 10^{-9}}$$ [1] $= 3.5 \times 10^{14}\,\text{m s}^{-2}$ [1]

3 a i The velocity increased from zero at a decreasing rate. [1]

ii The acceleration decreased gradually from a non-zero value. [1]

b Draw a tangent to the line at 0.5 s and measure the gradient of the tangent which should be $7.4\,\text{cm s}^{-2}$. [1] The acceleration at 0.5 s is therefore $7.4\,\text{cm s}^{-2}$ [1]

7.3

1 $u = 0$, $v = 28.0\,\text{m s}^{-1}$, $s = 170\,\text{m}$, $a = ?$, $t = ?$

    a To find a, use $v^2 = u^2 + 2as$. Rearranging this equation gives $2as = v^2 - u^2$

        Therefore, $a = \dfrac{v^2 - u^2}{2s} = \dfrac{28.0^2 - 0^2}{2 \times 170}$ [1] $= 2.3\,\text{m s}^{-2}$ [1]

    b Rearranging $s = \dfrac{(u+v)t}{2}$ to find t gives

        $t = \dfrac{2s}{u+v} = \dfrac{2 \times 170}{0 + 28.0}$ [1] $= 12.1\,\text{s}$ [1]

2 $u = 86\,\text{m s}^{-1}$, $v = 0$, $t = 28\,\text{s}$, $a = ?$, $s = ?$

    a To find s, use $s = \dfrac{(u+v)t}{2}$ to give

        $s = \dfrac{86 + 0}{2} \times 28$ [1] $= 1200\,\text{m}$ [1]

    b To find a, rearrange $v = u + at$ to give

        $a = \dfrac{v - u}{t} = \dfrac{0 - 86}{28}$ [1] $= -3.1\,\text{m s}^{-2}$ [1]

3 The displacement is given by the area under the curve. Each block of graph grid represents a displacement of 2.0 cm (because the height of each block represents a speed of $2.0\,\text{m s}^{-1}$ and the width of each block represents a time interval of 1.0 s). [1] Counting the number of grid blocks under the line from 0 to 2.5 s gives 21.0 ± 0.5 blocks so the displacement is 21.0 ± 0.5 cm. [1]

7.4

1 a $u = 0$, $v = ?$, $s = -1.21\,\text{m}$, $a = -9.81\,\text{m s}^{-2}$, $t = ?$

      i To find v, using $v^2 = u^2 + 2as$ gives
$v^2 = 0 + (2 \times -9.81 \times -1.21)$ [1] $= 23.7$
so $v = -4.87\,\text{m s}^{-1} = 4.9\,\text{m s}^{-1}$ to 2 significant figures. [1]

      ii To find t, rearranging $v = u + at$ gives
$t = \dfrac{v - u}{a} = \dfrac{-4.87 - 0}{-9.81}$ [1] $= 0.50\,\text{s}$ [1]

    b $u = ?$, $v = 0$ at maximum height, $s = +0.95\,\text{m}$, $a = -9.81\,\text{m s}^{-2}$, $t = ?$

        To find u, using $v^2 = u^2 + 2as$ gives $0 = u^2 + (2 \times -9.81 \times 0.95)$ [1] so $u^2 = (2 \times 9.81 \times 0.95) = 18.6$. Therefore, $u = +4.3\,\text{m s}^{-1}$ to 2 significant figures. [1]

2 $u = +2.5\,\text{m s}^{-1}$, $v = ?$, $s = -13\,\text{m}$, $a = -9.81\,\text{m s}^{-2}$, $t = ?$

    To find t, find v first then find t $\left(\text{rather than using the quadratic equation } s = ut + \dfrac{1}{2}at^2\right)$.

    Step 1: Using $v^2 = u^2 + 2as$ gives
$v^2 = 2.5^2 + (2 \times -9.81 \times -13)$ [1] so $v^2 = 261$.
Therefore, $v = -16.2\,\text{m s}^{-1}$. (− because the velocity is downwards) [1]

    Step 2: Rearranging $v = u + at$ gives
$t = \dfrac{v - u}{a} = \dfrac{-16.2 - 2.5}{-9.81}$ [1] $= 1.9\,\text{s}$ to 2 significant figures. [1]

3 a Dividing $s = ut + \dfrac{1}{2}at^2$ by t gives $\dfrac{s}{t} = u + \dfrac{1}{2}at$
Therefore, a graph of $\dfrac{s}{t}$ against t should give a straight line with a y-intercept equal to u and a gradient $\dfrac{1}{2}a$. [1]
As $u = 0$, the y-intercept should therefore be zero and the line should go through the origin. [1]

    b i A systematic error in a set of measurements is where there is a pattern (or a bias) in the measurements that causes them to differ from the true measurements. [1]

      ii A systematic error in s would give the same constant gradient and a non-zero y-intercept [1] whereas a systematic error in t would not give a constant gradient as well as a non-zero intercept. [1] Since the line is straight and does not pass through the origin, there is likely to have been a systematic error in the measurement of s. [1] and it would give a different gradient as well as a different y-intercept. [1]

7.5/7.6

1 a See Figure 1. A is at the end of the straight section of the line. B is where the line ends where the gradient is zero. [2]

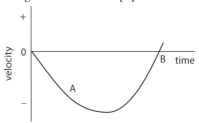

▲ Figure 1

    b The acceleration is constant and negative and equals $-g$ from O to A [1] and then it gradually decreases to zero [1] and then becomes a deceleration reducing the jumper's velocity to zero at B. [1]

2 a i $u = 0$, $v = -2.4\,\text{m s}^{-1}$ (− for downwards), $a = -9.81\,\text{m s}^{-2}$, $t = ?$

        To find t, rearranging $v = u + at$ gives
$t = \dfrac{v - u}{a} = \dfrac{-2.4 - 0}{-9.81}$ [1] $= 0.245\,\text{s} = 0.24\,\text{s}$ to 2 significant figures [1]

      ii See Figure 2 (straight line from the origin [1], velocity and time values marked at 0.245 s [1])

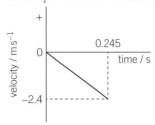

▲ Figure 2

    b Distance fallen = area between the line and the time axis in Figure 2
$= \dfrac{1}{2} \times 2.4 \times 0.245$ [1] $= 0.294\,\text{m}$ [1]
Note that the equation $v^2 = u^2 + 2as$ can be used to calculate the displacement ($= -0.294\,\text{m}$) which gives 0.29 m for the distance fallen to 2 significant figures.

3 a $a = +3.7\,\text{m s}^{-2}$, $u = 0$, $t = 30\,\text{s}$, $v = ?$, $s = ?$

        To calculate v, using $v = u + at$ gives
$v = 0 + 3.7 \times 30$ [1] $= 110\,\text{m s}^{-1}$ (2 s.f.) [1]

To calculate s, using $s = ut + \frac{1}{2}at^2$ gives

$s = 0 + 0.5 \times 3.7 \times 30^2$ [1] $= 1670$ (2 s.f.) m [1]

b i After the engines were switched off, $a = -9.81\,\text{m s}^{-2}$, $u = +111\,\text{m s}^{-1}$, and $v = 0$ at maximum height. To find t, rearranging $v = u + at$ gives $t = \dfrac{v - u}{a} = \dfrac{0 - 111}{-9.81}$ [1] $= 11$ s (2 s.f.) [1]

ii See Figure 3. The graph has two straight line sections OA and AB. [1] The gradient of section OA is $+3.7\,\text{m s}^{-2}$ and A is the point 30 s from zero where the velocity is $111\,\text{m s}^{-1}$.[1] The gradient of section AB is $-9.81\,\text{m s}^{-2}$ and B is the point where $v = 0$ at 41 s from the start. [1]

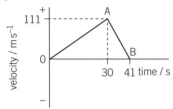

▲ **Figure 3**

c To calculate the velocity just before hitting the ground, consider the motion from 30 s after launch to the ground. For this section of the flight, $a = -9.81\,\text{m s}^{-2}$, $u = +111\,\text{m s}^{-1}$, and $s = -1665\,\text{m}$ (to the ground). Using the equation $v^2 = u^2 + 2as$ gives $v^2 = 111^2 + (2 \times -9.81 \times -1665)$ [1] $= 1.23 \times 10^4 + 3.23 \times 10^4 = 4.50 \times 10^4\,\text{m}^2\text{s}^{-2}$. [1] Therefore, $v = -210\,\text{m s}^{-1}$ (2 s.f.). (– for downwards) [1]

Note: The speed just before impact can also be calculated by adding the loss of potential energy per unit mass due to a height difference of 1665 m to the kinetic energy per unit mass at a speed of $111\,\text{m s}^{-1}$ to give the kinetic energy per unit mass and hence the speed just before impact.

7.7/7.8

1 a Consider the vertical motion: $u_y = 0$, $t = 3.1$ s, $a_y = -9.81\,\text{m s}^{-2}$ (– for downwards)

Using $y = \frac{1}{2}a_y t^2$ gives $y = \frac{1}{2} \times -9.81 \times 3.1^2 = -47$ m [1]

Therefore, the clifftop is 47 m above the sea. [1]

b Consider the horizontal motion: $u_x = 16\,\text{m s}^{-1}$, $a_x = 0$, $t = 3.1$ s

Therefore, the horizontal displacement $x = u_x t = 16 \times 3.1 = 50$ m [1]

c At $t = 3.1$ s, vertical component of velocity $v_y = u_x + a_y t = 0 - 9.81 \times 3.1$ [1] $= -30.4\,\text{m s}^{-1}$ (– for downwards) [1]

Therefore, impact speed $\sqrt{v_x^2 + v_y^2} = \sqrt{16^2 + 30.4^2} = 34\,\text{m s}^{-1}$ [1]

2 a Consider the vertical motion: $u_y = 0$, $t = 7.5$ s, $a_y = -9.81\,\text{m s}^{-2}$ (– for downwards)

Using $y = \frac{1}{2}a_y t^2$ gives $y = \frac{1}{2} \times -9.81 \times 7.5^2$ [1] $= -280$ m to 2 significant figures.

Therefore, the aircraft is 280 m above the sea. [1]

b Assuming the drag force is negligible, the horizontal component of the parcel's velocity will be equal to the aircraft's velocity so the parcel will be directly below the aircraft as it falls. [1]

If the drag force is not negligible, the horizontal component of the drag force will reduce the horizontal component of the parcel's velocity [1] so its horizontal displacement at the point of impact will be less than that of the aircraft and the horizontal position of the parcel will fall behind the plane's horizontal position. [1]

3 a Horizontal component $u_x = 29.0 \cos 15.0° = 28.0\,\text{m s}^{-1}$

Vertical component $u_y = 29.0 \sin 15.0° = 7.51\,\text{m s}^{-1}$ [1]

b Consider the vertical motion: $u_y = 7.51\,\text{m s}^{-1}$, $y = -2.60$ m, $a_y = -9.81\,\text{m s}^{-2}$

(– for downwards)

i To calculate the vertical component of velocity, v_y, using $v_y^2 = u_y^2 + 2a_y y$, gives:

$v_y^2 = 7.51^2 + (2 \times -9.81 \times -2.60)$ [1] $= 56.4 + 51.0 = 107.4\,\text{m}^2\text{s}^2$

Therefore, $v_y = -10.4\,\text{m s}^{-1}$ [1]

ii To calculate the time taken t rearranging $v_y = u_y + a_y t$ gives:

$t = \dfrac{v_y - u_y}{a_y} = \dfrac{-10.4 - 7.51}{-9.81}$ [1] $= 1.83$ s [1]

c Consider the horizontal motion: $u_x = 28.0\,\text{m s}^{-1}$, $a_x = 0$, $t = 1.83$ s

Therefore, the horizontal displacement $x = u_x t = 28.0 \times 1.83 = 51.2$ m [1]

8.1

1 a $u = 0$, $v = 12\,\text{m s}^{-1}$, $t = 50$ s

acceleration $a = \dfrac{v - u}{t} = \dfrac{12 - 0}{50} = 0.24\,\text{m s}^{-2}$ [1]

b $F = ma = 26\,000 \times 0.24 = 6200\,\text{N}$ (2 s.f.) [1]

c Weight $= mg$

Ratio of resultant force:weight $= \dfrac{ma}{mg} = \dfrac{0.24}{9.81}$ $= 0.024$ to 2 significant figures [1]

2 a $u = 62\,\text{m s}^{-1}$, $v = 0$, $s = 1600$ m

Rearranging $v^2 = u^2 + 2as$ gives

$a = \dfrac{v^2 - u^2}{2s} = \dfrac{0 - 62^2}{2 \times 1600}$ [1] $= -1.2\,\text{m s}^{-2}$ [1]

b $F = ma = 4000 \times -1.2 = -4800\,\text{N}$

Therefore, the decelerating force $= 4800\,\text{N}$ [1]

3 a $F = 6000\,\text{N}$, $m = 0.027$ kg

acceleration $a = \dfrac{F}{m} = \dfrac{6000}{0.027} = 2.2 \times 10^5\,\text{m s}^{-2}$ to 2 significant figures [1]

b i Average speed $= \dfrac{\text{distance}}{\text{time}} = \dfrac{310\,\text{m}}{4.2\,\text{s}}$

$(= 73.8\,\text{m s}^{-1}) = 74\,\text{m s}^{-1}$ to 2 significant figures. [1]

ii During the contact time t assume the ball accelerated from rest to $74\,\text{m s}^{-1}$ with an acceleration of $2.2 \times 10^5\,\text{m s}^{-2}$.

Rearranging $v = u + at$ gives $t = \dfrac{v - u}{a} = \dfrac{74 - 0}{2.2 \times 10^5}$ [1] $= 3.4 \times 10^{-4}$ s [1]

c 1 The golf ball is assumed to have negligible vertical motion which is not the case as the ball leaves the golf club with a vertical component of velocity as well as a horizontal component. [1] The average speed calculation assumes the vertical component of its velocity is negligible. The actual contact time is therefore longer than the calculated value in part **b**. [1]

2 Air resistance is assumed to be negligible which is not the case as air resistance on the golf ball reduces its speed and its range. [1] Without air resistance the average speed would have been greater which would therefore give a longer contact time. [1]

8.2

1 a Weight = mg = 1050 × 9.81 = 10 300 N [1]

b $T - mg = ma$ where T is the thrust [1]

$T = mg + ma$ = 10 300 + (1050 × 4.80) = 15 300 N [1]

2 a At constant velocity, the tension T in the cable is equal and opposite to the weight.

Therefore, $T = mg$ = 1200 × 9.81 = 11 800 N [1]

b $T - mg = ma$ [1] therefore $T = mg + ma$

Since the lift is accelerating downwards, then $a = -0.50\,\text{m s}^{-2}$. [1]

Therefore, $T = mg + ma$ = 11 800 + (1200 × −0.50) = 11 200 N [1]

3 a $u = 0$, $s = 5.0\,\text{m}$, $t = 3.0\,\text{s}$

To calculate a, substitute $u = 0$ into $s = ut + \frac{1}{2}at^2$

to give $s = \frac{1}{2}at^2$ and rearrange the equation to give

$a = \dfrac{2s}{t^2} = \dfrac{2 \times 5.0}{3.0^2}$ [1] = 1.1 m s^{-2} (2 s.f.) [1]

b i Component of weight acting down the slope
= $mg\sin\theta$ = 36 × 9.81 × sin 20 = 120 N [1]

ii The resultant force on the skateboarder
= $mg\sin\theta - F_0$ where F_0 is the frictional force.

Therefore, $mg\sin\theta - F_0 = ma$ [1] hence
$F_0 = mg\sin\theta - ma$ = 120 − (36 × 1.1) = 80 N [1]

8.3

1 a Terminal speed = $\dfrac{\text{distance}}{\text{time}} = \dfrac{0.32\,\text{m}}{7.1\,\text{s}}$ = 0.045 m s^{-1} [1]

b At terminal speed, the drag force = weight of the object [1]

= mg = 0.045 kg × 9.81 N kg^{-1} = 0.44 N [1]

2 a See Figure 1 (OW correct [1]. Beyond W, decrease at decreasing gradient to constant speed [1])

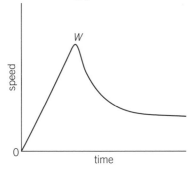

▲ **Figure 1**

b After the ball is released, until it reaches the water surface, it has a constant acceleration equal to g because the air resistance on it is negligible and it accelerates due to the force of gravity on it. When it enters the water, because it is moving fast, it experiences a drag force which is greater than its weight [1] so the resultant force on it is in the opposite direction to its velocity and it decelerates and slows down. [1] As it slows down the drag force on it decreases so the resultant force on it decreases [1] and its deceleration decreases gradually to zero [1] when the drag force is equal to its weight and it moves at terminal speed. [1]

3 a maximum acceleration =
$\dfrac{\text{maximum engine force}}{\text{mass}} = \dfrac{3800}{22000}$ = 0.17 m s^{-2} [1]

b At top speed, the total resistive force = its maximum engine force = 3800 N [1]

8.4

1 a i Thinking distance = initial speed × reaction time = 20 m s^{-1} × 0.70 s = 14 m (2 s.f.) [1]

ii $u = 20\,\text{m s}^{-1}$, $v = 0$, $a = -5.1\,\text{m s}^{-2}$, braking distance = s to be calculated

To calculate s, rearrange $v^2 = u^2 + 2as$ to give
$s = \dfrac{v^2 - u^2}{2a} = \dfrac{0 - 20^2}{2 \times -5.1}$ [1] = 39 m (2 s.f.) [1]

b Distance from the stopped vehicle to the pedestrian = 58.0 − (14.0 + 39.2) = 4.8 m [1]

2 a $u = 28\,\text{m s}^{-1}$, $v = 0$, $s = 71\,\text{m}$

To calculate s, rearrange $v^2 = u^2 + 2as$ to give
$a = \dfrac{v^2 - u^2}{2s} = \dfrac{0 - 28^2}{2 \times 71}$ [1] = −5.5 m s^{-2} [1]

b Resultant force = mass × acceleration
= 15 000 × −5.5 = 83 000 N

The braking force is 83 000 N [1]

(assuming the resultant force is due to the frictional force only).

3 a Stopping distance is the sum of the thinking distance and the braking distance. The braking force on a vehicle on a wet road must be reduced compared with on a dry road otherwise the vehicle will skid. [1] So the braking distance for a vehicle on a wet road is longer than on a dry road for the same initial speed. Therefore, the stopping distance is longer than on a dry road. [1]

b Braking force = 0.57mg

Deceleration = $\dfrac{\text{braking force}}{\text{mass}} = \dfrac{0.57mg}{m}$ [1] = 0.57g
= 0.57 × 9.81 = 5.59 m s^{-2} [1]

$u = 31\,\text{m s}^{-1}$, $v = 0$, $a = -5.59\,\text{m s}^{-2}$ (− for deceleration)

To calculate s, rearrange $v^2 = u^2 + 2as$ to give
$s = \dfrac{v^2 - u^2}{2a} = \dfrac{0 - 31^2}{2 \times -5.59}$ [1] = 86 m [1]

8.5

1 $u = 12\,\text{m s}^{-1}$, $v = 18\,\text{m s}^{-1}$, $s = 3.0\,\text{m}$

impact time $t = \dfrac{2s}{u + v} = \dfrac{2 \times 3.0}{12 + 18}$ [1] = 0.20 s [1]

acceleration $a = \dfrac{v - u}{t} = \dfrac{18 - 12}{0.20}$ = 30 m s^{-2} [1]

impact force $F = ma$ = 1100 × 30 = 33 000 N [1]

2 $u = 3.0\,\mathrm{m\,s^{-1}}$, $v = 0$, $t = 0.40\,\mathrm{s}$

acceleration $a = \dfrac{v - u}{t} = \dfrac{0 - 3}{0.40} = -7.5\,\mathrm{m\,s^{-2}}$ [1]

impact force $F = ma = 900 \times 7.5 = 6800\,\mathrm{N}$ [1]

3 a $u = 23\,\mathrm{m\,s^{-1}}$, $v = 0$, $s = 4.8 + 0.5 = 5.3\,\mathrm{m}$

impact time $t = \dfrac{2s}{u + v} = \dfrac{2 \times 5.3}{23 + 0}$ [1] $= 0.46$ s [1]

acceleration $a = \dfrac{v - u}{t} = \dfrac{0 - 23}{0.46} = -50\,\mathrm{m\,s^{-2}}$ [1]

impact force $F = ma = 62 \times 50 = 3100\,\mathrm{N}$ [1]

b As the car braked, the passenger would move at a speed of $23\,\mathrm{m\,s^{-1}}$ until she collided with the inside of the car. [1] The impact time in this collision would be much shorter than the impact time when she was wearing the seat belt. [1] So the force of the impact without the seat belt on would be much larger than the force on her with the seat belt on. [1]

9.1/9.2

1 Change of momentum $= \Delta(mv) = 25\,000 \times 160 = 4.0 \times 10^6\,\mathrm{N\,s}$ [1]

Force $= \dfrac{\Delta(mv)}{\Delta t} = \dfrac{4.0 \times 10^6}{58} = 69\,000$ N [1]

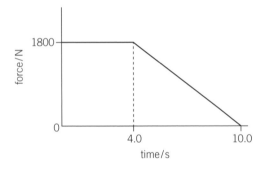

▲ **Figure 1**

2 a See Figure 1 [1]

b Change of momentum = area under line

$= (1800\,\mathrm{N} \times 4.0\,\mathrm{s}) + (0.5 \times 1800\,\mathrm{N} \times 6.0\,\mathrm{s})$
$= 12\,600\,\mathrm{N\,s}$ [1]

Change of velocity $= \dfrac{\text{change of momentum}}{\text{mass}}$

$= \dfrac{12\,600}{1300} = 9.7\,\mathrm{m\,s^{-1}}$ [1]

Final velocity $= 20 - 9.7 = 10.3\,\mathrm{m\,s^{-1}}$ [1]

3 Change of momentum = final momentum − initial momentum

$= (1500 \times 6.5) - (1500 \times 2.0) = 6750\,\mathrm{N\,s}$ [1]

Force $= \dfrac{\text{change of momentum}}{\text{contact time}} = \dfrac{6750\ \mathrm{Ns}}{0.55\ \mathrm{s}}$
$= 12\,000\,\mathrm{N}$ [1]

4 The normal component of its momentum is $+ mu \cos\theta$ before the impact and $-mu\cos\theta$ after the impact.

Change of momentum $= (-mu\cos\theta) - (mu\cos\theta)$
$= -2\,mu\cos\theta$ [1]

$= -2 \times 5.0 \times 10^{-26}\,\mathrm{kg} \times 420\,\mathrm{m\,s^{-1}} \times \cos 60$
$= 2.1 \times 10^{-23}\,\mathrm{kg\,m\,s^{-1}}$ [1]

Force $= \dfrac{\text{change of momentum}}{\text{contact time}} = \dfrac{2.1 \times 10^{-23}\ \mathrm{kg\,m\,s^{-1}}}{0.22 \times 10^{-9}\ \mathrm{s}}$
$= 9.5 \times 10^{-14}\,\mathrm{N}$ [1]

9.3/9.4/9.5

1 Let m = mass of B.

Total initial momentum $= (1.50 \times 0.35) + (m \times -0.25) = 0.525 - 0.25m$ [1]

Total final momentum $= (m + 1.50) \times 0.050 = 0.050m + 0.075$ [1]

Therefore, $0.050m + 0.075 = 0.525 - 0.25m$ [1]

Rearranging this equation gives:

$0.25m + 0.05\,\mathrm{m} = 0.525 - 0.075$ which gives

$0.30\,\mathrm{m} = 0.450$ so $m = \dfrac{0.450}{0.30} = 1.5\,\mathrm{kg}$ [1]

2 Let v = the velocity of Y after the collision.

Total initial momentum $= (0.60 \times 3.0)$
$= 1.80\,\mathrm{kg\,m\,s^{-1}}$ [1]

Total final momentum $= (0.60 \times 0.80) + (0.40 \times v)$
$= 0.48 + 0.40v$ [1]

Therefore, $0.40v + 0.48 = 1.80$

Rearranging this equation gives $0.40v = 1.80 - 0.48$
$= 1.32$ [1]

Therefore, $v = \dfrac{1.32}{0.40} = 3.3\,\mathrm{m\,s^{-1}}$ in the same direction as the direction of A initially. [1]

3 a Let v = the velocity of P after the collision.

Total initial momentum $= (3000 \times 1.2) + (2000 \times -2.5) = -1400\,\mathrm{kg\,m\,s^{-1}}$ [1]

Total final momentum $= (3000 \times v) + (2000 \times 0.50) = 3000v + 1000$ [1]

Therefore, $3000v + 1000 = -1400$ [1]

Rearranging this equation gives $3000v = -2400$

Therefore, $v = \dfrac{-2400}{3000} = -0.80\,\mathrm{m\,s^{-1}}$

The velocity of P after the collision $= 0.80\,\mathrm{m\,s^{-1}}$ in the opposite direction to its initial direction. [1]

b Kinetic energy E_K of P before the collision

$= \dfrac{1}{2} \times 3000 \times 1.2^2 = 2160\,\mathrm{J}$

E_K of Q before the collision $= \dfrac{1}{2} \times 2000 \times -2.5^2$
$= 6250\,\mathrm{J}$

Total E_K before collision $= 2160 + 6250 = 8410\,\mathrm{J}$ [1]

E_K of P after the collision $= \dfrac{1}{2} \times 3000 \times -0.80^2$
$= 960\,\mathrm{J}$

E_K of Q after the collision $= \dfrac{1}{2} \times 2000 \times 0.5^2$
$= 250\,\mathrm{J}$

Total E_K after collision $= 960 + 250 = 1210\,\mathrm{J}$ [1]

Total E_K after collision < total E_K before the collision, so the collision is inelastic. [1]

4 a Let V = recoil velocity of the gun.

Total final momentum $= (2.4 \times 150) + (900 \times V)$
$= 360 + 900V$

Total initial momentum $= 0$ [1]

Therefore, $360 + 900V = 0$ which gives $V = \dfrac{-360}{900}$
$= -0.40\,\mathrm{m\,s^{-1}}$

The recoil velocity of the gun $= 0.40\,\mathrm{m\,s^{-1}}$ [1]

b Kinetic energy of shell $= \dfrac{1}{2} \times 2.4 \times 150^2 = 27\,000\,\mathrm{J}$

Kinetic energy of gun $= \dfrac{1}{2} \times 900 \times -0.40^2 = 72\,\mathrm{J}$ [1]

Kinetic energy of gun/total kinetic energy
= 72/(27 000 + 72) = 0.0027

Therefore, kinetic energy of the gun as a percentage of the total kinetic energy
= 0.0027 × 100% = 0.27% [1]

10.1/10.2

1 a Work done = $Fs\cos\theta$ = 14 × 3.0 cos 60 = 21 J [1]

  b $W = \frac{1}{2}F\Delta L$ = 0.5 × 20 × 0.45 = 4.5 J [1]

2 a KE at ground = initial KE + loss of GPE

    $= \frac{1}{2}mu^2 + mg\,\Delta h$

    = (0.5 × 0.048 × 16.0²) + (0.048 × 9.81 × 23.0) [1]

    = 6.1 + 10.8 = 16.9 J [1]

  b Let v = speed just before impact so $\frac{1}{2}mv^2 = E_K$ (= 16.9 J)

    Rearranging to find v gives

    $v = \sqrt{\dfrac{2E_K}{m}} = \sqrt{\dfrac{2 \times 16.9}{0.048}}$ [1] = 28 m s⁻¹ (2 s.f.) [1]

3 a Loss of potential energy = $mg\,\Delta h$ = 80 × 9.81 × 40 = 31 J [1]

  b Gain of kinetic energy = $\frac{1}{2}mv^2$ = 0.5 × 80 × 14² = 7.8 kJ [1]

  c Work done to overcome resistive forces = loss of GPE − gain of KE = 31.4 − 7.8 = 23.6 kJ [1]

    average resistive force =

    $\dfrac{\text{work done}}{\text{distance moved}} = \dfrac{23.6\text{ kJ}}{560\text{ m}} = 42$ N [1]

10.3/10.4

1 a Rearranging power $P = Fv$ where F is the force and v is the velocity gives

    $F = \dfrac{P}{v} = \dfrac{240000\text{ W}}{31\text{ m s}^{-1}}$ [1] = 7.7 kN [1]

  b Useful energy transferred per second = 240 kJ s⁻¹ = 37% of energy per second supplied

    Energy supplied per second = 240 000 ÷ 0.37 = 650 kJ s⁻¹ [1]

    Energy wasted per second = 650 − 240 = 410 kJ s⁻¹ [1]

2 a Power $P = kv^3$ where k is the constant of proportionality

    Therefore, $k = \dfrac{P}{v^3}$ [1] $= \dfrac{1.4}{12^3}$ MW (m s⁻¹)⁻³ [1]

    Hence, at $v = 8$ m s⁻¹, $P = kv^3 = \dfrac{1.4}{12^3} \times 8.0^3 = 0.41$ MW [1]

  b Kinetic energy/second of the wind

    $= \dfrac{\text{output power}}{\text{efficiency}} = \dfrac{1.4\text{ MW}}{0.48}$ [1] = 2.9 MW [1]

3 Mass of water passing through per second= vol/sec × density = 52 × 1000 = 52 000 kg s⁻¹ [1]

  Loss of potential energy each second = $mg\,\Delta h$ (where m = 52 000 kg)

  = 52000 × 9.81 × 350 = 179 MJ s⁻¹ [1]

  Output power = 28% of 179 MJ s⁻¹ = 50 MW [1]

11.1

1 a Volume = 20 × 10⁻³ × 60 × 10⁻² × 75 × 10⁻² = 9.0 × 10⁻³ m³ [1]

  b Density = $\dfrac{\text{mass}}{\text{volume}} = \dfrac{22.4\text{ kg}}{9.0 \times 10^{-3}\text{ m}^3}$ = 2500 kg m⁻³ (2 s.f.) [1]

2 a Mass of liquid = 248 g − 85 g = 163 g = 0.163 kg [1]

  b Depth H of liquid in tin = 85 mm

    Volume of liquid = $\frac{1}{4}\pi d^2 \times H$ = 0.25 × π × (50 × 10⁻³)² × 85 × 10⁻³ = 1.67 × 10⁻⁴ m³ [1]

  c Density = $\dfrac{\text{mass}}{\text{volume}} = \dfrac{0.163\text{ kg}}{1.67 \times 10^{-4}\text{ m}^3}$ = 980 kg m⁻³ [1]

3 a i Volume of the wire = $\frac{1}{4}\pi d^2 \times L$ = 0.25 × π × (0.220 × 10⁻³)² × 2415 × 10⁻³ [1]

      = 9.18 × 10⁻⁸ m³ [1]

   ii Density of metal = $\dfrac{\text{mass}}{\text{volume}} = \dfrac{0.730 \times 10^{-3}\text{ kg}}{9.18 \times 10^{-8}\text{ m}^3}$

      = 7950 kg m⁻³ [1]

  b % uncertainties: mass = $\dfrac{0.005}{0.730} \times 100\%$ = 0.7%;

    length = $\dfrac{5}{2415} \times 100\%$ = 0.2%; [1]

    diameter = $\dfrac{0.005}{0.220} \times 100\%$ = 2.27% therefore

    % uncertainty in the area of cross-section

    = 2 × 2.27% = 4.54%. [1] Total % uncertainty = 0.7 + 0.2 + 4.54 = 5.44% [1]

    Therefore, uncertainty in the density value = 5.44% of 7950 = 430 kg m⁻³ [1]

11.2

1 a Rearranging $F = k\Delta L$ gives $\Delta L = \dfrac{F}{k} = \dfrac{8.0\text{ N}}{40\text{ Nm}^{-1}}$ = 0.20 m [1]

  b Energy stored = $\frac{1}{2}F\Delta L$ = 0.5 × 8.0 × 0.20 = 0.80 J [1]

2 a i Since the springs are identical and in parallel, the tension T in each spring is the same and equal to half the weight supported. So T = 0.80 N. [1]

   ii Extension ΔL of each spring = 274 − 250 = 24 mm [1]

      Tension T in each spring = $k\Delta L$ where k is the spring constant of each spring.

      Hence, $k = \dfrac{T}{\Delta L} = \dfrac{0.80\text{ N}}{0.024\text{ m}}$ [1] = 33.3(3) N m⁻¹ = 33 N m⁻¹ to 2 significant figures [1]

  b Total energy stored = $\frac{1}{2}k\Delta L^2$ where the effective spring constant k = 2 × 33.3(3) = 66.7 N m⁻¹ and ΔL = 0.024 m. [1]

    Therefore, the total energy stored = $\frac{1}{2}$ × 66.7 × 0.024² [1] = 0.019 J. [1]

3 a The tension T in each spring is the same because they are in series and is equal to the weight supported. Hence, T = 1.6 N. [1]

  b For P, ΔL = 280 − 250 = 30 mm. [1] Therefore, $k_P = \dfrac{T}{\Delta L} = \dfrac{1.6\text{ N}}{0.030\text{ m}}$ = 53 N m⁻¹ [1]

    For Q, ΔL = 300 − 250 = 50 mm. [1] Therefore, $k_P = \dfrac{T}{\Delta L} = \dfrac{1.6\text{ N}}{0.050\text{ m}}$ = 32 N m⁻¹ [1]

11.3/11.4

1 Area of cross-section $A = \frac{1}{4}\pi d^2 = 0.25\pi \times$
$(0.26 \times 10^{-3})^2 = 5.3 \times 10^{-8}\,\mathrm{m^2}$ [1]

$E = \frac{FL}{A\Delta L} = \frac{48 \times 1.524}{5.3 \times 10^{-8} \times 6.6 \times 10^{-3}}$ [1] $= 2.1 \times 10^{11}\,\mathrm{Pa}$ [1]

2 a Area of cross-section $A = \frac{1}{4}\pi d^2 = 0.25\pi \times$
$(0.22 \times 10^{-3})^2 = 3.8 \times 10^{-8}\,\mathrm{m^2}$ [1]

Rearranging
$E = \frac{FL}{A\Delta L}$ gives $\Delta L = \frac{FL}{AE} = \frac{32 \times 1.500}{3.8 \times 10^{-8} \times 2.1 \times 10^{11}}$ [1]
$= 6.0 \times 10^{-3}\,\mathrm{m}$ [1]

b Elastic energy stored $= \frac{1}{2}T\Delta L = 0.5 \times 32 \times$
$6.0 \times 10^{-3} = 0.096\,\mathrm{J}$ [1]

3 a Rearranging $E = \frac{F}{A\Delta L}$ where F is the compressive
force gives stress
$= \frac{F}{A} = \frac{E\Delta L}{L} = \frac{2.1 \times 10^{11} \times 0.13 \times 10^{-3}}{0.050}$ [1]
$= 5.5 \times 10^8\,\mathrm{Pa}$ [1]

b Compressive force = stress $\times A = 5.5 \times 10^8 \times$
3.4×10^{-4} [1] $= 187\,\mathrm{kN}$ [1]
Elastic energy stored $= \frac{1}{2}F\Delta L = 0.5 \times 187 \times 10^3 \times$
$0.13 \times 10^{-3} = 12\,\mathrm{J}$ to 2 significant figures [1]

12.1/12.2

1 a $\Delta Q = I\Delta t = 1.2 \times 10^{-3} \times 300\,\mathrm{s} = 0.36\,\mathrm{C}$ [1]

b no. of electrons $= \frac{\Delta Q}{e} = \frac{0.36\,\mathrm{C}}{1.60 \times 10^{-19}\,\mathrm{C}}$
$= 2.3 \times 10^{18}$ (2 s.f.) [1]

2 a Current $I = \frac{P}{V} = \frac{800\,\mathrm{W}}{230\,\mathrm{V}} = 3.5\,\mathrm{A}$ (2 s.f.) [1]

b Energy transferred in 1 minute $= Pt$
$= 800\,\mathrm{W} \times 60\,\mathrm{s} = 48\,000\,\mathrm{J}$ [1]

3 a Rearranging $\Delta Q = I\Delta t$ gives
$\Delta t = \frac{\Delta Q}{I} = \frac{3.6 \times 10^6\,\mathrm{C}}{13} = 2.8 \times 10^5\,\mathrm{s}$ (2 s.f.) [1]

b Rearranging $V = \frac{W}{Q}$ gives energy delivered
$= QV = 3.6 \times 10^6\,\mathrm{C} \times 12\,\mathrm{V} = 4.3 \times 10^7\,\mathrm{J}$ (2 s.f.) [1]
(or energy delivered $= IV\Delta t = 13\,\mathrm{A} \times 12\,\mathrm{V} \times$
$2.77 \times 10^5\,\mathrm{s} = 4.3 \times 10^7\,\mathrm{J}$ (2 s.f.))

12.3

1 a $V = IR = 0.970\,\mathrm{A} \times 0.156\,\Omega = 0.151\,\mathrm{V}$ [1]

b Energy per second dissipated $= IV$
$= 0.970\,\mathrm{A} \times 0.151\,\mathrm{V} = 0.146\,\mathrm{W}$ (3 s.f.) [1]

2 Area of cross-section of wire
$A = 35.1 \times 10^{-3}\,\mathrm{m} \times 0.62 \times 10^{-3}\,\mathrm{m} = 2.18 \times 10^{-5}\,\mathrm{m^2}$ [1]

Resistivity $\rho = \frac{RA}{L} = \frac{0.156 \times 2.18 \times 10^{-5}\,\mathrm{m^2}}{0.382\,\mathrm{m}}$
$= 8.90 \times 10^{-6}\,\Omega\,\mathrm{m}$ [1]

3 a Area of cross-section $= \frac{1}{4}\pi d^2$
$= 0.25\pi \times (1.28 \times 10^{-3})^2 = 1.29 \times 10^{-6}\,\mathrm{m^2}$ [1]

Rearranging,
$\rho = \frac{RA}{L}$ gives $R = \frac{\rho L}{A} = \frac{1.70 \times 10^{-8}\,\mathrm{m} \times 25.0\,\mathrm{m}}{1.29 \times 10^{-6}\,\mathrm{m^2}}$
[1]

$= 0.329\,\Omega$ [1] (to 2 significant figures)

b Energy dissipated per second $P = I^2R = 13.0^2$
$\times 0.329\,\Omega = 55.6\,\mathrm{W}$ [1]

12.4

1 a See Figure 1. [1]

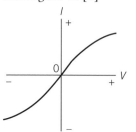

▲ Figure 1

b The resistance increases as the current increases
in either direction. [1] The increase of resistance
is because the filament becomes hotter as the
current increases and the positive ions in the
filament vibrate more. [1] The conduction
electrons need to do more work because they
collide more with the positive ions [1] so a
greater potential difference is needed to maintain
the same current. [1]

2 a i $I = \frac{V}{R} = \frac{4.5\,\mathrm{V}}{100\,000} = 4.5 \times 10^{-5}\,\mathrm{A}$ [1]

ii $I = \frac{V}{R} = \frac{4.5\,\mathrm{V}}{400} = 1.1 \times 10^{-2}\,\mathrm{A}$ (2 s.f.) [1]

b The number of charge carriers in the LDR
increases if the incident light intensity is
increased. [1] Therefore, there are fewer charge
carriers in the LDR in darkness than in daylight
so the current for the same pd is smaller in
darkness than in daylight. Hence, the resistance
in darkness is larger than in daylight. [1]

3 a See Figure 1. [1]

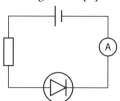

▲ Figure 1

b When the diode conducts, the pd across it does
not change when the current increases. The pd
across the resistor is therefore unchanged if the
current changes. [1] The ammeter reading would
decrease because the resistor has the same pd
across its terminals and its resistance is greater so
the current in the circuit would be less. [1]

13.1

1 a See Figure 1. [1]

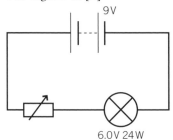

▲ Figure 1

b i 4.0 A [1]

ii Pd across variable resistor = 9.0 − 6.0 = 3.0 V [1]

iii Power supplied by battery = IV = 4.0 A × 9.0 V = 36 W [1]

2 a See Figure 2. [1]

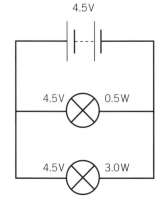

▲ Figure 2

b $I = \dfrac{P}{V} = \dfrac{0.5\,\text{W}}{4.5\,\text{V}} = 0.11\,\text{A}$ for the 0.5 W torchbulb [1]

$I = \dfrac{P}{V} = \dfrac{3.0\,\text{W}}{4.5\,\text{V}} = 0.67\,\text{A}$ for the 3.0 W torchbulb [1]

c Battery current = 0.11 + 0.67 = 0.78 A [1] hence energy per second supplied by the battery = IV = 0.78 A × 4.5 V = 3.5 W [1]

Energy supplied per second to the two torchbulbs = 3.0 W + 0.5 W = 3.5 W = energy per second supplied by the battery. [1]

3 a See Figure 3. [1]

▲ Figure 3

b i Pd across 1.5 kΩ resistor (= current × resistance) = 2.0 mA × 1.5 kΩ = 3.0 V [1]

ii Pd across resistor R = 9.0 − 3.0 = 6.0 V [1]

iii Resistance of $R = \dfrac{V}{I} = \dfrac{6.0\,\text{V}}{2.0\,\text{mA}} = 3.0\,\text{kΩ}$ [1]

13.2

1 a See Figure 1. [1]

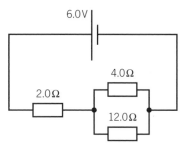

▲ Figure 1

b i Resistance of the 2 parallel resistors

$= \left(\dfrac{1}{4.0} + \dfrac{1}{12.0}\right)^{-1} = 3.0\,\text{Ω}$ [1]

Total circuit resistance = 3.0 + 2.0 = 5.0 Ω [1]

ii Battery current = $\dfrac{\text{cell emf}}{\text{total circuit resistance}}$

$= \dfrac{6.0\,\text{V}}{5.0} = 1.2\,\text{A}$ [1]

c Pd across 2.0 Ω resistor = IR = 1.2 × 2.0 = 2.4 V therefore pd across parallel resistors

= 6.0 − 2.4 = 3.6 V [1]

Power supplied to 4.0 Ω resistor = $\dfrac{V^2}{R} = \dfrac{3.6^2}{4.0} = 3.2$ W (2 s.f.) [1]

2 a See Figure 2 (all correct [2], correct circuit with incorrect labels [1])

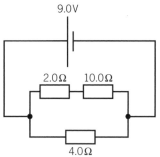

▲ Figure 2

b i Resistance of the 2 series resistors = 2.0 Ω + 10.0 Ω = 12.0 Ω [1]

Total circuit resistance = $\left(\dfrac{1}{4.0} + \dfrac{1}{12.0}\right)^{-1}$ = 3.0 Ω [1]

ii Battery current = $\dfrac{\text{cell emf}}{\text{total circuit resistance}}$

$= \dfrac{9.0\,\text{V}}{3.0\,\text{Ω}} = 3.0\,\text{A}$ [1]

c Power supplied to 4.0 Ω resistor

$= \dfrac{V^2}{R} = \dfrac{9.0^2}{4.0} = 20.3$ W [1]

Current through the 2.0 Ω and 10.0 Ω resistors

$= \dfrac{V}{R} = \dfrac{9.0\,\text{V}}{12.0\,\text{Ω}} = 0.75$ A

Power supplied to 2.0 Ω resistor = I^2R = 0.75^2 × 2.0 = 1.1 W (2 s.f.) [1]

Power supplied to 10.0 Ω resistor = I^2R = 0.75^2 × 10.0 = 5.6 W (2 s.f.) [1]

Note: The power supplied by the battery = IV = 3.0 A × 9.0 V = 27 W. This is equal to the total power supplied to the resistors (= 20.25 W + 1.125 W + 5.625 W as required by conservation of energy).

3 $P = 3000\,\text{W}$, $V = 230\,\text{V}$. Rearranging $P = \dfrac{V^2}{R}$ gives
$R = \dfrac{V^2}{P} = \dfrac{230^2}{3000}$ [1] $= 18$ (2 s.f.) [1]

13.3

1 a Total resistance $= 4.5 + 0.5 = 5.0\,\Omega$ [1]

b Battery current $= \dfrac{V}{R} = \dfrac{9.0\,\text{V}}{5.0} = 1.8\,\text{A}$ [1]

c Pd across cell terminals $= IR = 1.8\,\text{A} \times 4.5\,\Omega$
$= 8.1\,\text{V}$ [1] (or $\varepsilon - Ir = 9.0 - (1.8 \times 0.5) = 8.1\,\text{V}$)

2 a Current $= \dfrac{\text{emf}}{\text{total circuit resistance}} = \dfrac{2.0}{3.5 + 0.5}$ [1]
$= 0.50\,\text{A}$ [1]

b Power delivered to 3.5 Ω resistor $= I^2 R$
$= 0.50^2 \times 3.5 = 0.88\,\text{W}$ (2 s.f.) [1]

c Power wasted $= I^2 r = 0.50^2 \times 0.5 = 0.13\,\text{W}$ (2 s.f.) [1]

3 a Using $\varepsilon = V + Ir$ with $V = 1.05\,\text{V}$ and $I = 0.25\,\text{A}$ gives
$\varepsilon = 1.05 + 0.25r$ [1]

Using $\varepsilon = V + Ir$ with $V = 1.35\,\text{V}$ and $I = 0.10\,\text{A}$ gives
$\varepsilon = 1.35 + 0.10r$ [1]

Therefore, $1.35 + 0.10r = 1.05 + 0.25r$ which gives
$0.25r - 0.10r = 1.35 - 1.05$

Therefore, $r = \dfrac{1.35 - 1.05}{0.25 - 0.10} = \dfrac{0.30}{0.15} = 2.0\,\Omega$ [1]

b Using $\varepsilon = V + Ir$ with $V = 1.05\,\text{V}$, $I = 0.25\,\text{A}$ and
$r = 2.0\,\Omega$ gives:
$\varepsilon = 1.05 + (0.25 \times 2.0\,\Omega) = 1.55\,\text{V}$ [1]

13.4

1 a See Figure 1. [1]

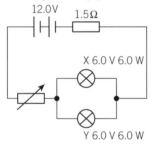

▲ Figure 1

b i Current in each lamp $= \dfrac{P}{V} = \dfrac{6.0\,\text{W}}{6.0\,\text{V}} = 1.0\,\text{A}$ [1]

ii Battery current $I = I_X + I_Y = 1.0 + 1.0 = 2.0\,\text{A}$ [1]

iii Battery emf = pd across variable resistor + pd
across lamps + pd across internal resistance
Pd across internal resistance $= Ir = 2.0 \times 1.5$
$= 3.0\,\text{V}$ [1]
Pd across lamps $= 6.0\,\text{V}$
Therefore, pd across variable resistor
$= 12.0\,\text{V} - 6.0\,\text{V} - 3.0\,\text{V} = 3.0\,\text{V}$ [1]

13.5

1 a See Figure 1. [1]

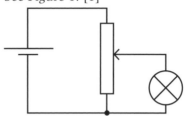

▲ Figure 1

b A potential divider can be adjusted to give a range
of pd from zero to a maximum equal to the pd
of the supply voltage. Therefore, the brightness
of the lamp can be adjusted from zero to a
maximum by adjusting the potential divider. [1]
A variable resistor in series with the lamp gives a
range of current between a minimum when the
variable resistor has maximum resistance to a
maximum current when the variable resistor has
zero resistance. [1] Therefore, the brightness of a
lamp cannot be reduced to zero using a variable
resistor. [1]

2 a i Total circuit resistance $= 8.0\,\Omega + 12.0\,\Omega$
$= 20.0\,\Omega$ [1]

Current in each resistor
$= \dfrac{\text{battery pd}}{\text{total circuit resistance}} = \dfrac{6.0\,\text{V}}{20.0\,\Omega} = 0.30\,\text{A}$ [1]

ii Let V_1 = pd across the $8.0\,\Omega$ resistor and
V_2 = pd across the $12.0\,\Omega$ resistor. The battery
pd of 6.0 V is shared between the $8.0\,\Omega$ resistor
and the $12.0\,\Omega$ resistor in proportion to their
resistances. Therefore, $\dfrac{V_2}{V_1} = \dfrac{12.0}{8.0} = 1.5$ so
$V_2 = 1.5\,V_1$. [1]
So the battery pd $= V_1 + V_2 = V_1 + 1.5\,V_1$
$= 2.5\,V_1$
Hence, $2.5\,V_1 = 6.0\,\text{V}$ so $V_1 = \dfrac{6.0}{2.5} = 2.4\,\text{V}$ and
$V_2 = 1.5\,V_1 = 1.5 \times 2.4 = 3.6\,\text{V}$ [1]
(OR pd across each resistor = current × resistance)
Therefore, pd across $8.0\,\Omega$ resistor
$= 0.30\,\text{A} \times 8.0\,\Omega = 2.4\,\text{V}$ [1]
(and pd across $12.0\,\Omega$ resistor
$= 0.30\,\text{A} \times 12.0\,\Omega = 3.6\,\text{V}$)

b i The total circuit resistance $= 16.0 + 8.0$
$= 24.0\,\Omega$. The ratio of the pd across the 16 Ω
resistor to the cell pd $= 16.0/24.0 = 0.667$. [1]
Therefore, the pd across the 16 Ω resistor $=$
$0.667 \times$ the cell pd $= 0.667 \times 4.5\,\text{V} = 3.0\,\text{V}$. [1]

(Alternative method: current in the circuit
= 4.5 V / (16.0 Ω + 8.0 Ω) = 0.188 A

Therefore, pd across 16 Ω resistor
= 0.188 A × 16.0 Ω = 3.0 V)

ii The total circuit resistance = 6.0 + 8.0 = 14.0 Ω.
The ratio of the pd across the 6 Ω resistor to
the cell pd = 6.0/14.0 = 0.429. [1] Therefore,
the pd across the 6 Ω resistor = 0.429 × the cell
pd = 0.429 × 4.5 V = 1.9(3) V. [1]

(Alternative method: current in the circuit
= 4.5 V / (6.0 Ω + 8.0 Ω) = 0.321 A

Therefore, pd across 6 Ω resistor
= 0.321 A × 6.0 Ω = 1.9(3) V)

3 a See Figure 2. [1]

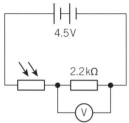

▲ **Figure 2**

b i Pd across LDR = 4.5 V − 1.2 V = 3.3 V [1]

ii Current through 2.2 kΩ resistor
$= \frac{V}{R} = \frac{1.2\ V}{2.2\ k\Omega} = 0.55$ mA [1]

LDR resistance $= \frac{V}{I} = \frac{3.3\ V}{0.55\ mA} = 6.0$ kΩ [1]

c When the LDR is exposed to daylight, its resistance
decreases so the share of the battery pd across the
LDR decreases. [1] Therefore, the pd across the
resistor increases. [1]

For reference

Useful data for AS Physics
Data

Fundamental constants and values

quantity	symbol	value	units
speed of light in vacuo	c	3.00×10^8	$m\,s^{-1}$
charge of electron	e	-1.60×10^{-19}	C
the Planck constant	h	6.63×10^{-34}	J s
the Avogadro constant	N_A	6.02×10^{23}	mol^{-1}
molar gas constant	R	8.31	$J\,K^{-1}\,mol^{-1}$
the Boltzmann constant	k	1.38×10^{-23}	$J\,K^{-1}$
electron rest mass	m_e	9.11×10^{-31}	kg
electron charge/mass ratio	e/m_e	1.76×10^{11}	$C\,kg^{-1}$
proton rest mass	m_p	$1.67(3) \times 10^{-27}$	kg
proton charge/mass ratio	e/m_p	9.58×10^{7}	$C\,kg^{-1}$
neutron rest mass	m_n	$1.67(5) \times 10^{-27}$	kg
gravitational field strength	g	9.81	$N\,kg^{-1}$
acceleration due to gravity	g	9.81	$m\,s^{-2}$
atomic mass unit	u	1.661×10^{-27}	kg

AS equations
Particle physics

Fundamental particles

class	name	symbol	rest energy / MeV
photon	photon	γ	0
lepton	neutrino	ν_e	0
		ν_μ	0
	electron	$e^\pm$	0.510999
	muon	$\mu^\pm$	105.659
mesons	π meson	$\pi^\pm$	139.576
		π^0	134.972
	K meson	$K^\pm$	493.821
		K^0	497.762
baryons	proton	p	938.257
	neutron	n	939.551

Properties of quarks
Antiparticles have opposite signs

type	charge	baryon number	strangeness
u	$+\frac{2}{3}e$	$+\frac{1}{3}$	0
d	$-\frac{1}{3}e$	$+\frac{1}{3}$	0
s	$-\frac{1}{3}e$	$+\frac{1}{3}$	-1

Properties of leptons

lepton	lepton number
particles: $e^-, \nu_e\,; \mu^-, \nu_\mu\,; \tau^-, \nu_\tau$	$+1$
antiparticles: $e^+, \overline{\nu}_e\,; \mu^+, \overline{\nu}_\mu\,; \tau^+, \overline{\nu}_\tau$	-1

Geometrical equations

arc length $= r\theta$

circumference of circle $= 2\pi r$

area of circle $= \pi r^2$

surface area of cylinder $= 2\pi rh$

volume of cylinder $= \pi r^2 h$

area of sphere $= 4\pi r^2$

volume of sphere $= \frac{4}{3}\pi r^3$

Photons and energy levels

photon energy $\qquad\qquad E = hf = \dfrac{hc}{\lambda}$

photoelectric effect $\qquad hf = \varphi + E_{Kmax}$

energy levels $\qquad\qquad hf = E_1 - E_2$

de Broglie wavelength $\quad \lambda = \dfrac{h}{p} = \dfrac{h}{mv}$

Electricity

current and pd $\qquad\qquad I = \dfrac{\Delta Q}{\Delta t}$

$$V = \dfrac{W}{Q}$$

$$R = \dfrac{V}{I}$$

emf $\qquad\qquad\qquad\qquad \varepsilon = \dfrac{W}{Q}$

$$\varepsilon = IR + Ir$$

resistors in series $\qquad\quad R = R_1 + R_2 + R_3 + \dots$

resistors in parallel $\qquad \dfrac{1}{R} = \dfrac{1}{R_1} + \dfrac{1}{R_2} + \dfrac{1}{R_3} + \dots$

resistivity $\qquad\qquad\qquad \rho = \dfrac{RA}{L}$

power $\qquad\qquad\qquad\quad P = VI = I^2 R = \dfrac{V^2}{R}$

Mechanics

Moments $\qquad\qquad\qquad$ moment $= Fd$

velocity and acceleration $\quad v = \dfrac{\Delta s}{\Delta t}; \ a = \dfrac{\Delta v}{\Delta t}$

equations of motion $\qquad\quad v = u + at$

$$s = \dfrac{(u + v)t}{2}$$

$$v^2 = u^2 + 2as$$

$$s = ut + \dfrac{1}{2}at^2$$

force $\qquad\qquad\qquad\qquad F = ma$

$$F = -\dfrac{\Delta(mv)}{\Delta t}$$

work, energy and power $\qquad W = Fs \cos\theta$

$$E_K = \dfrac{1}{2}mv^2$$

$$\Delta E_P = mg\Delta h$$

$$P = \dfrac{\Delta W}{\Delta t} \qquad P = Fv$$

efficiency of a machine $= \dfrac{\text{useful output power}}{\text{input power}}$

Materials

density $\qquad\qquad\qquad\qquad \rho = \dfrac{m}{v}$

Hooke's law $\qquad\qquad\qquad F = k\Delta L$

Young modulus $\qquad\quad = \dfrac{\text{tensile stress}}{\text{tensile strain}} = \dfrac{FL}{A\,\Delta L}$

energy stored $\qquad\qquad E = \dfrac{1}{2}k\,\Delta L^2 = \dfrac{1}{2}F\Delta L$

Waves

wave speed $\qquad\qquad\qquad c = f\lambda$

period $\qquad\qquad\qquad\qquad T = \dfrac{1}{f}$

fringe spacing $\qquad\qquad\quad w = \dfrac{\lambda D}{s}$

diffraction grating $\qquad\quad d \sin\theta = n\lambda$

refractive index of a substance s, $n_s = \dfrac{c}{c_s}$

For two different substances of refractive indices n_1 and n_2,

law of refraction $\qquad\qquad n_1 \sin\theta_1 = n_2 \sin\theta_2$

critical angle $\qquad\qquad\quad \sin\theta_c = \dfrac{n_2}{n_1}$ for $n_1 > n_2$

For a stretched string, first harmonic frequency

$$f_0 = \dfrac{1}{2l}\sqrt{\dfrac{T}{m}}$$